Amartya Sen's Capability Approach and Social Justice in Education

Edited by
Melanie Walker and Elaine Unterhalter

AMARTYA SEN'S CAPABILITY APPROACH AND SOCIAL JUSTICE IN EDUCATION
Copyright © Melanie Walker and Elaine Unterhalter, 2007.

First published in hardcover in 2007 by
PALGRAVE MACMILLAN® in the United States – a division of
St. Martin's Press LLC, 175 Fifth Avenue, New York, NY 10010.

Where this book is distributed in the UK, Europe and the rest of the world,
this is by Palgrave Macmillan, a division of Macmillan Publishers Limited,
registered in England, company number 785998, of Houndmills,
Basingstoke, Hampshire RG21 6XS.

Palgrave Macmillan is the global academic imprint of the above companies
and has companies and representatives throughout the world.

Palgrave® and Macmillan® are registered trademarks in the United States,
the United Kingdom, Europe and other countries.

ISBN: 978–0–230–10459–4

Library of Congress Cataloging-in-Publication Data

Amartya Sen's capability approach and social justice in education / Melanie
Walker and Elaine Unterhalter (editors.)
        p. cm.
    Includes bibliographical references and index.
    ISBN 1–4039–7504–3 (alk. paper)
    1. Educational sociology.   2. Education—Economic aspects.
3. Learning Ability.   4. Social Justice.   5. Sen, Amartya Kumar.
I. Walker, Melanie.   II. Unterhalter, Elaine.
    LC191.A56 2007
    370.11′5—dc22

                                                        2007060032

A catalogue record of the book is available from the British Library.

Design by MPS Limited, A Macmillan Company

First PALGRAVE MACMILLAN paperback edition: July 2010

10 9 8 7 6 5 4 3 2 1

Printed in the United States of America.

Transferred to Digital Printing in 2010

Amartya Sen's Capability Approach and Social Justice in Education

WITHDRÁWN

In memory of
Professor Terence McLaughlin, 1949–2006

Contents

List of Tables and Figures

Tables

Figures

Acknowledgments

This book arises both from the Capability and Education seminar network we established in December 2003, and from the international conferences of the Human Development and Capability Association (HDCA) that have been held annually since 2001. Contributors and participants in the seminar network and the conferences have been a source of intellectual challenge and engagement, and many of them have chapters in this book. The education network has been supported from the outset by Flavio Comim and David Bridges from the Von Hugel Institute, St Edmund's College, University of Cambridge, and we are deeply grateful for the base they provided in establishing and sustaining the network in the first two years. Severine Deneulin, who worked at the Von Hugel Institute from 2003 to 2005, has actively participated in the network and her substantial contribution to the depth of discussion merits particular note. Terence McLaughlin was an early and continuing enthusiast for considering the education issues associated with the capability approach, and his untimely death in March 2006 meant the loss of a careful and critical commentator on this project. In completing it we want to record our deep appreciation for his warmth, insight, and openness.

Work within the multidisciplinary environment of the HDCA has provided an invaluable opportunity to consider how education is similar to as well as different from other areas of social policy and how work in education links with debates in philosophy, politics, and economics. Our particular thanks to Sabina Alkire, Enrica Chiappero Marinetti, Jean Luc Dubois, Des Gasper, Mozaffar Quizalbash, and Ingrid Robeyns for many useful discussions on situating our work in education in a wider context.

We both warmly acknowledge the contribution of our graduate students at the Institute of Education and the University of Sheffield; they have been a source of critical questions when confronted with the capability approach that has helped shape the ideas in this book. We have been delighted when some of them have chosen to continue their inquiries into the capability approach in relation to higher education, gender equity or education, and international development for their own doctoral dissertations. Former colleagues at the University of Sheffield, colleagues at the

Institute of Education, University of London, and collaborators on a range of other research projects in other institutions have been enthusiastic and supportive of our respective capability projects: Ann Marie Bathmaker, Sue Webb, Gareth Parry, Dan Goodley, Jennifer Lavia, Monica McLean, Vivienne Bozalek, Brenda Leibowitz, Gloria D'Allba, Deborah Gaitskell, Robert Cowen, Chris Yates, Madeleine Arnot, Harry Brighouse, Sheila Aikman, Amy North, and Linda Chisholm.

Anjali Kothari and Janet Raynor did an invaluable job in helping to prepare the manuscript for publication, while Amanda Johnson, the commissioning editor at Palgrave, has been both patient and enthusiastic.

Our final thanks go to our respective partners and families for patiently living through yet another book project.

Melanie Walker and Elaine Unterhalter

Notes on Contributors

Richard Bates is professor of education in the Faculty of Education at Deakin University, Australia. He is president of the Australian Teacher Education Association, a past president of the Australian Association for Researchers in Education and the Australian Council of Deans of Education, and a fellow of the Australian College of Education and the Australian Council for Educational Administration. He is currently writing about morals and markets, public education, ethics and administration, the impact of educational research, social justice and the aesthetics of educational administration, and teacher education.

Mario Biggeri is associate professor in development economics at the University of Florence, and vice-coordinator of the PhD program in Politics and Economics of Developing Countries. He has worked as consultant for UNICEF Innocenti Research Centre for the research project "Home Based Workers and Child Labour in Manufacturing Processes in South and Southeast Asia" (2000–2002), for the ILO/UNICEF/World Bank Project *Understanding Children's Work* (UCW) (2003), and for various NGOs. He is coauthor of a book on international aid and coeditor of two other books. He has published papers in *Journal of International Development, Journal of Chinese and Business Studies, World Development,* and *Journal of Human Development*.

Harry Brighouse is professor of philosophy at the University of Wisconsin-Madison. He works on the foundations of liberal theory, and is especially interested in the place of education and the family in liberalism. He is currently working on the significance of positionality for egalitarian justice, on development theory and liberal egalitarianism, on the fair distribution of power, and on whether parents have rights over their children. He has also written extensively about education policy in both scholarly journals and the British national press. He is the author of *School Choice and Social Justice* (Oxford University Press, 2000). He is coeditor, with Randall Curren, Mitja Sardoc, and Janez Justin, of the journal *Theory and Research in Education*.

Sandra S. Butler is professor and interim director of the School of Social Work at the University of Maine. Her research focuses on financial security issues for women across the life span. She has published *Middle-aged, Female, and Homeless: The Stories of a Forgotten Group* (Garland Press, 1994), *Shut Out: Low Income Mothers and Higher Education in Post-Welfare America* (SUNY Press, 2004), and *Gerontological Social Work in Small Towns and Rural Communities* (Haworth Press, 2004), as well as numerous articles and book chapters on the restrictions of current welfare policy on low-income women seeking to access higher education, income security, rural gerontology, and the impact of health and welfare policy on the lives of women.

Luisa S. Deprez is professor in the department of sociology and the women's studies program at the University of Southern Maine. Her scholarly interests and publications center on the broad arenas of social welfare policy including the politics of policymaking; the impact of ideology and public opinion in policy; citizenship; poverty; and women, welfare, and higher education. She has published *The Family Support Act of 1988: A Case Study of Welfare Policy in the 1980s* (Edwin Mellen Press, 2002) and *Shut Out: Low Income Mothers and Higher Education in Post-Welfare America* (SUNY Press, 2004), as well as numerous articles and book chapters about the restrictions of current welfare policy on low-income women seeking to access higher education.

Pedro Flores-Crespo is lecturer in the Research Institute for the Development of Education at the Universidad Iberoamericana, Mexico. His research focuses on education and human development, and education policy; currently, he is working on ethnicity, identity, and educational achievement. He has published various articles on inequality, human capabilities, and the analysis of policy process. His most recent book is *Desarrollo como Libertad en América Latina: Fundamentos y Aplicaciones* (forthcoming), edited with Mathias Nebel. He was the network coordinator of the Human Development and Capability Association from 2007 to 2009.

Janet Raynor has worked for over nine years in a number of education programs in Bangladesh, with an increasing focus on gender issues on education, and with studies linked to girls' education and the development of capabilities in Bangladesh. She has published a number of chapters in edited collections based on this work. She is currently working in Vietnam on an ADB/Government of Vietnam project focussing on education in the most disadvantaged regions, with an emphasis on girls' education.

Lorella Terzi is a philosopher of education with training in political and moral philosophy. She is the author of *Justice and Equality in Education: a Capability Perspective on Disability and Special Educational Needs* (Continuum, 2008), which innovatively applies Amartya Sen's capability approach to questions of provision for children with disabilities and special educational needs. She is a Reader at Roehampton University, UK.

Elaine Unterhalter is Professor of Education and International Development at the Institute of Education, University of London, UK. She contributes to the MA course on Education, Gender, and International Development and to a range of other post-graduate teaching, and works on a number of research projects on gender and education in Africa. Her research interests include feminism, global social justice, and education policy and politics. Her recent books include *Gender, schooling and global social justice* (2007); *Gender equality, HIV and AIDS: Challenges for the education sector* (with Sheila Aikman and Tania Boler, 2008); *Towards equality? Gender in South African schools during the HIV and AIDS epidemic* (with Robert Morrell, Debbie Epstein, Lebo Moletsane, and Devia Bhana, 2009). *Global Inequalities and Higher Education* (co-edited with Vincent Carpentier) will be published in 2010 by Palgrave.

Rosie Vaughan has recently completed a PhD at the Faculty of Education, University of Cambridge, UK. Her research interests are in gender, education and capabilities, and the measurement of educational inequalities. She is currently conducting her doctoral research on the relationship between international organizations and the Indian government in the promotion of girls' education.

Melanie Walker is Professor of Higher Education and Director of Research in the School of Education at the University of Nottingham, UK, and Extraordinary Professor at the University of the Western Cape, South Africa. She is director of a PhD program in Higher Education, and contributes to teaching the MA in Higher Education and the Professional Doctorate. Her teaching and research interests focus on human development and global higher education policy; professional education, graduate capabilities, and university contributions to poverty reduction; and pedagogies and social justice. She is co-editor of the *Journal of Human Development and Capabilities*. Her recent books include *Higher Education Pedagogies: a capabilities approach* (2006); and *The Routledge Doctoral Supervisor's Companion* and *The Routledge Doctoral Student's Companion*, both edited with Pat Thomson (2010).

1

The Capability Approach: Its Potential for Work in Education

Melanie Walker and Elaine Unterhalter

Amartya Sen is one of the key thinkers and commentators of the late twentieth and early twenty-first centuries. Influential as a Nobel Prize–winning economist and a political philosopher, Sen is a key contributor to identifying, detailing, and campaigning against forms of global inequality. A major theme of his work is how to evaluate human well-being. His ideas on evaluation, equality, freedom, and rights stand at the center of the capability approach, which is generally associated with his name, having its origins in lectures he delivered in the late 1970s (Sen 1980). The capability approach rests on a critique of other approaches to thinking about human well-being in welfare economics and political philosophy, which are concerned with commodities, a standard of living, and justice as fairness. The capability approach challenges elements of these formulations and entails a consideration of evaluation, policy, and action that has had considerable impact both within the disciplines in which it emerged and within development theory concerned with analyses of poverty. An emerging literature has considered the implications of the capability approach for education; this book brings together conceptual and empirical writings on this theme.

Before considering the significance of the capability approach for discussions in education, some of its core ideas and the specific terms associated with these need explaining.

What are Capabilities?

Sen defines a capability as "a person's ability to do valuable acts or reach valuable states of being; [it] represents the alternative combinations of things a person is able to do or be" (Sen 1993, p. 30). Thus, capabilities are opportunities or freedoms to achieve what an individual reflectively considers valuable. The significance of this idea rests on its contrast with other ideas concerning how we decide what is just or fair in the distribution of resources. For example, some ideas about distribution rest on what an outsider determines is best to create maximum opportunities or achieve appropriate outcomes for, say, different kinds of schools or students. The problem is often phrased in terms of what forms of curriculum, teaching, school management, and learning resources will yield the education achievements, such as examination results or skill sets, that an economy needs. Sometimes the question is posed in terms of how learners can acquire appropriate knowledge of history or religion to act as full members of a particular group to which they are deemed to belong. In both these instances the emphasis is on what kinds of inputs (ideas, teachers, learning materials) will shape particular opportunities to achieve desired outcomes (economic growth or social solidarity). Ideas influenced by utilitarianism pose this in terms of outcomes deemed the best result for the largest number, for example, the numbers of people who will benefit nationally and internationally from growth in an economy or the numbers of people who will draw together through practices of religious or cultural belonging.

The capability approach critiques this way of posing and solving questions of evaluation. Its central tenet is that in evaluations one must look at each person not as a means to economic growth or social stability but as an end. We must evaluate freedoms for people to be able to make decisions they value and work to remove obstacles to those freedoms, that is, expand people's capabilities. Importantly, while the capability approach regards each human being as an end, it is not an individualistic framework concerned with libertarian notions of self-actualization above all other goods. Rather it embraces "ethical individualism" (Robeyns 2005, p. 108), a normative approach that stresses that actions should be judged by their effects on individual human beings and that individuals are the "primary objects of moral concern"(Brighouse and Swift 2003, p. 358).

Evaluation is thus not simply a response to what particular individuals want or say they want. Designing policy only to respond to what people want could mean that a government might use up nearly all the education budget for a country to provide resources for the small number of children of vocal parents who want schooling only in lavish buildings, with one-to-one tuition, leaving very meager resources for the majority of children whose parents want the best education the government can afford. Evaluating capabilities, rather than resources or outcomes, shifts the axis of analysis to establishing and evaluating the conditions that enable individuals to take decisions based on what they have reason to value. These conditions will vary in different contexts, but the approach sets out to be sensitive to human diversity; complex social relations; a sense of reciprocity between people; appreciation that people can reflect reasonably on what they value for themselves and others; and a concern to equalize, not opportunities or outcomes, but rather capabilities.

Equality of What?

The capability approach thus offers a broad normative framework to conceptualize and evaluate individual well-being and social arrangements in any particular context or society. It is not a complete theory of justice, but it does deal with questions of the balance between freedoms and equality that have characterized work on social justice since the late eighteenth century. Sen (1980) asks the core question, "Equality of what?" As he explains, all egalitarian theories that have stood the test of time pose the issue of equality of something, for example, of income, welfare levels, rights, or liberties. In education this question emerges in philosophical and sociological work on how to theorize and analyze the provision of equivalent learning opportunities (e.g., Brighouse 2000; Ball 2003).

The choice of the space in which to assess equality determines what equality we prioritize. We could prioritize equalizing the income of every adult in a country and thus place income equality in the space of evaluation. Sen argues that what we should equalize is not resources, for example, a strict ratio of teachers to pupils, or a certain amount of expenditure per capita on each pupil, and not outcomes, for example, that every child leaves school with a particular qualification. He writes that what should be equalized are human capabilities, that is, what people are able to be and to do.

Crucial to this is the process for people to come to decisions about what they have reason to value in and from education, or any other aspect of social action. Thus the expansion of human capability involves "the freedoms [people] actually enjoy to choose the lives that they have reason to value" (Sen 1992, p. 81). People should be able to make choices that matter to them for a valuable life. The notion of capability "is essentially one of freedom—the range of options a person has in deciding what kind of life to lead" (Dreze and Sen 1995, p. 11). Capabilities might then also be explained as "actions one values doing or approaches to living one's values" (Unterhalter 2003b, p. 666).

In other words, when we evaluate social (and educational) arrangements against a criterion of justice and considerations of equalities, it is people's capabilities that must guide the evaluation rather than how much money, education resources, or qualifications they are able to command. Resources are the means, but not the intrinsic ends, of human well-being. In education we evaluate capabilities in and through education as complex outcomes. Evaluation is concerned with a dynamic relationship between opportunity and outcome, or capabilities and functionings, as described below.

Capabilities and Functionings

A second core idea in the capability approach is the distinction between capabilities and functionings (Sen 1980). Functionings are achieved outcomes. Reading, talking to children, taking part in the social life of a community through attending a meeting about a school, being calm, are all functionings. Capabilities are the potential to achieve these functionings, for example, having been taught to read; having books or newspapers available to read; living in a society where adults of your class, gender, or race are permitted to talk to children and attend meetings at a school; and having the conditions that will develop calm (not too long a working day or too many anxieties). The difference between a capability and functioning is one between an opportunity to achieve and the actual achievement, between potential and outcome.

This distinction is very important because evaluating only functionings or outcomes can give too little information about how well people are doing. Some cases may look as though the same functionings have been achieved but behind these equal outcomes may lie very different stories, and it is the difference that is germane to thinking about justice and equality. Here are two examples of apparently equitable educational outcomes.

Two young women complete a degree in literature at the same English university. One, from a middle-class, reasonably affluent background and a good school, wished to experience university before working in her father's business as a trainee manager. An outstanding degree result was not required. Nonetheless she coped well with the academic demands of her course, having been suitably prepared by her school. She enjoyed the challenges entailed in contesting ideas in seminars. The second young woman, from a working-class background and a struggling inner city state school, despite significant academic ability, struggled to fit in and make friends among her middle-class peers. The teaching methods at her school had not prepared her well for higher education. Contestation over ideas in class undermined her confidence and made her anxious and unwilling to advance an opinion. She nonetheless worked hard, desperate to get excellent grades, but her lack of confidence meant she blamed herself for her struggles and was reluctant to approach her tutors for help with work. Both students obtained second-class passes. This shows that similar functionings mask very different capability sets.

Here is another example. Two 13-year-old girls in Kenya participating in an international study of learning achievements fail mathematics. For one, despite attending a well-equipped school in Nairobi with qualified and motivated teachers offering ample learning support and safe learning environment, a major reason for failure was her decision to spend less time on mathematics and more time with friends in the drama club and other leisure activities. For the other, from a school in Wajir, one of Kenya's poorest districts, despite her interest in mathematics and schoolwork generally, her results were largely due to the lack of a mathematics teacher at her school. The subject was taught by an English specialist. Private after-school tuition was available, but her parents could not afford this for all their children. They decided to prioritize their son and required their daughter to perform housework and childcare. She had little time to prepare for the examinations.

If we look only at functioning in these two examples—performance in examinations—we see equal (if regrettable, in the case of mathematics) outcomes. But while the functionings of the students are the same, their capabilities are different. The capability approach requires that we do not simply evaluate the functionings but the real freedom or opportunities each student had available to choose and to achieve what she valued. Our evaluation of equality must then take account of freedom in opportunities as much as observed choices. The capability approach, therefore, offers a method to evaluate real educational advantage, and equally to identify disadvantage, marginalization, and exclusion. In the approach, individual capabilities to undertake valued and valuable activities constitute an indispensable and central part of the relevant informational base of evaluation.

Freedoms and Agency

Freedoms and agency are the third group of concepts that require some further elucidation as they are central to the capability approach. People are understood to be active participants in development, rather than passive spectators. Agency here is taken to mean that each person is a dignified and responsible human being who shapes her or his own life in the light of goals that matter, rather than simply being shaped or instructed how to think.

These goals might not necessarily make an individual happier or more comfortable, but they are reached through reasoned reflection.

In education we are the agents of our own learning, the agents or instruments of the learning (or failure to learn) of others, and the recipients of others' agency. Agency deserves our attention in the way it potentially enables us to imagine and act toward new ways of being. For Sen (1999), to be actively involved in shaping one's own life and having opportunities to reflect on this is critical for positive social change. Agency is intrinsically important for individual freedom, he argues, but also instrumental for collective action and democratic participation. We exercise our agency individually and in cooperation with others, and through educational opportunities and appropriate processes we might learn to do both. Agency is also a key dimension of human well-being, as Alkire (2002) argues, and is further able to expand or advance our well-being in ways we deem worthwhile. Like learning, thinking of oneself as an agent whose actions and contributions count in the world of education does not happen overnight. It is a process of both being and becoming. By embracing agency in and through our education practices, we open the possibility to interrupt a pervasive relationship in education that tends to link learners' origins and outcomes.

In considering agency we need to ask if different learners are recognized socially and educationally as having equal claims on resources and opportunities. Nussbaum (2000) points to the difficulty with "adapted preferences." As she explains, our subjective preferences and choices are shaped and informed or deformed by society and public policy. Unequal social and political circumstances (both in matters of redistribution and recognition) lead to unequal chances and unequal capacities to choose. These external (material as well as cultural) circumstances "affect the inner lives of people: what they hope for, what they love, what they fear, as well as what they are able to do" (Nussbaum 2000, p. 31). Our choices are deeply shaped by the structure of opportunities available to us so that a disadvantaged group comes to accept its status within the hierarchy as correct even when it involves a denial of opportunities. Such adapted preferences can limit individual aspirations and hopes for the future. We adjust our hopes to our probabilities, even if these are not in our best interests. In turn, our agency and well-being are diminished rather than enhanced, even if we do not see it in this way. Although not describing education, Sen captures the problem when he writes,

> The destitute thrown into beggary, the vulnerable landless laborer precariously surviving at the end of subsistence, the over-worked domestic servant working around the clock, the subdued and subjugated housewife reconciled to her role and her fate, all tend to come to terms with their respective predicaments.
>
> (Sen 1985, p. 15, quoted in Comim 2003, p. 7)

In the field of education many schoolteachers recognize the child who finds it difficult to work and says, "I am a low achiever," while in higher education the importance of building confidence and motivation as much as encouraging knowledge and understanding are acknowledged. Why is the capability approach significant for social justice in education? Evaluating capabilities rather than functionings is a significant contribution to discussions of social justice in education, including increasing attention to notions of agency and identity. Debate in this field can be grouped into two broad streams. One has focused on how schools reproduce inequalities and social injustices through maldistribution and silencing (Bourdieu and Passeron 1977; Bowles and Gintis 1976; Ball 2003; Bowles and Gintis 2002; Kwesiga 2002; Aikman 1999) A second considers how conditions in schools or other learning sites offer resources or conditions through which learners can contest or change inequalities (Stromquist 1998; McLeod 2005; Lynch and Baker 2005; Brighouse 2002). The first group of writers, who see school as contributing to continuing injustices and inequalities, tend to pay attention to outcomes and functionings. The second group of writers, who are concerned with how schools might contribute to developing equalities and conditions for social justice, analyze aspects of the transformative space of schooling even if it is imperfectly realized. They are thus concerned with some of the process dimensions of justice and the same range of issues as the capability approach. For the first group of authors, larger structures or systems, associated with class or race or gender, will always form the conditions of justice. The role of evaluation is to document these effects of race, class, and gender. For the second group, there is a space of human action in which these structures are contested and where equalities in education can be formed. Evaluating and documenting these processes and what constrains or enlarges them sociologically or philosophically is part of the social justice project. The capability approach adds to these concerns with substantive equality and aspects of personal action, a particular emphasis on the capability-functioning relationship, and a consideration of human diversity and variability that is inherent in the approach at a very high level of abstraction and not just in application.

Education in the Capability Approach

There is some difficulty in separating out education or schooling analytically or politically (as we do in this book), given the multidimensionality of human development, capabilities, and education. Education is central to the capability approach. Sen (1992, p. 44), for example, identifies education as one of "a relatively small number of centrally important beings and doings that are crucial to well-being." Nussbaum (1997, 2002, 2004, 2006), in her work on higher education in the United States and India and on schooling in India, has discussed the importance of education for women's empowerment and the importance of public education as crucial to democratic societies. She identifies three key capabilities associated with education: first, critical thinking or "the examined life"; second, the ideal of the world citizen; and third, the development of the narrative imagination (see Nussbaum 2006; 1997).

In both Sen's and Nussbaum's works, education is in itself a basic capability that affects the development and expansion of other capabilities (but see Terzi, chapter 2, who disagrees that education can itself be a basic capability, and Vaughan, chapter 6). Having the opportunity for education and the development of an education capability expands human freedoms. Not having education harms human development and choosing and having a full life. Education, argues Sen (1999), fulfills an *instrumental social role* in that critical literacy, for example, fosters public debate and dialogue about social and political arrangements. It has an *instrumental process role* by expanding the people one comes into contact with, broadening our horizons. Finally, it has an *empowering and distributive role* in facilitating the ability of the disadvantaged, marginalized, and excluded to organize politically. It has *redistributive* effects between social groups, households, and within families. Overall, education contributes to *interpersonal* effects where people are able to use the benefits of education to help others and hence contribute to the social good and democratic freedoms. In short, for Sen, "education" is an unqualified good for human capability expansion and human freedom.

The approach thus leads us to ask questions such as: Are valued capabilities distributed fairly in and through education? Do some people get more opportunities to convert their resources into capabilities than others, and if so who, how, and why? In short, it means taking up the crucial importance Sen (1999) allocates to education in the formation of human capabilities.

Education, Individual Development, and Social Arrangements

In educational studies there has been much concern with the relationship between educational and social inequalities. Drawing on the capability approach while analyzing education provides a useful vocabulary to engage with these issues. Capability and functionings depend on individual circumstances, the relations a person has with others, and social conditions and contexts within which potential options (or freedoms) can be achieved. Individual agency, Sen insists, depends also on social and eco-nomic arrangements (for example, the level of provision of education or health care) and on political and civil rights, so that Sen comments that there is "a deep complementarity between individual agency and social arrangements" and that individual freedom is "a social commitment" (Sen 1999, pp. xi–xii).

The capability approach foregrounds the basic heterogeneity of human beings as a fundamental aspect of educational equality and connects indi-vidual biographies and social and collective arrangements. Social context and social relations can enlarge or constrain individual capabilities for education and in education; personal and relational differences set condi-tions for capabilities. Thus Sen's (1992, 1999) metric of equality includes both personal evaluation and interpersonal variation, as well as individual and social arrangements. Equality thus depends on aspects of personal responsibility for actions, the differences between people that will constrain or facilitate particular outcomes, and the social relations that secure those differences. If we think of the two Kenyan schoolgirls discussed earlier, each differed in their individual and social circumstances with regard to school provision and family support, but in addition each also differed in the extent to which she took and was able to take responsibility for studying mathematics, and in her capacity to understand mathematical ideas and use these in the context of the examination set. At issue is that "goods and services are not the only means to people's capabilities" (Robeyns 2005, p. 99). As Robeyns (2005) explains, both material and nonmaterial circumstances shape the opportunities that we have and the choices that we make.

Individual functionings are influenced by a person's relative advantages in society and enhanced by enabling public and policy environments, for example, a gender-equity or antidiscrimination policy regarding disabili-ties. Sen emphasizes that "being free to live the way one would like may be enormously helped by the choices of others, and it would be a mistake to think of achievements only in terms of active choice by oneself" (Sen 1993, p. 44). Social opportunities and social norms expand human agency or diminish it. Social norms in nonideal contexts (everyday, real life) construct disadvantages, even where public resources might be equally distributed. A learner's opportunities may be significantly helped by the choices of others—good teachers, productive peer relationships, enabling public policy. Sen, therefore, integrates the personal and the macrosocial in securing and expanding intrapersonal and interpersonal freedoms.

Conversion Factors and Learner Diversity

Sen's (1992) concept of conversion is crucial in making the capability approach sensitive to the impact and effect of social arrangements and social relations on individual lives. He argues that equalizing the ownership of resources "need not equalize the substantive freedoms enjoyed by different persons, since there can be significant variations in the conversion of resources and primary goods into freedoms" (p. 33). Learners differ in intersecting dimensions. These include personal differences such as enthusiasm for academic study or artistic ability; environmental differences, such as wealth or whether children live in a society with a history of education inequalities such as the UK or greater equalities such as Sweden; and social differences, for example, the extent to which race, ethnic, or gender differences are salient with regard to the experience of education. There is nothing inherently unequal about difference or the intersection of differences, but differences can become inequalities (Terzi 2005). For example, a learner might value the capability for voice, but finds herself silenced in a classroom through particular social arrangements of power and privilege. To convert her capability into a functioning she needs social arrangements that are sensitive to her ways of expressing herself and give her opportunities for this. This requires particular forms of classroom pedagogy and management and the resources for this that are not only fixed assets such as staff, but also training, cultures of concern with learners' difference, and the capacity to put this care into practice. Terzi's (2005) example of a visually impaired learner highlights how the concept of conversion works, showing that being a visually impaired learner is a disadvantage when specific resources such as Braille texts are not provided, or where the physical environment is not appropriately designed. Thus, Terzi argues that disability is relational both to individual impairments and to the design and provision of educational arrangements. Attention to conversion signals concerns with what is needed to realize particular functionings.

Nussbaum (2000) tackles the issue of individual capability and social arrangements with her concept of "combined capabilities." These comprise our "internal capabilities," which Nussbaum (2000, p. 84) explains as the "developed states of the individual herself that are, so far as the person herself is concerned, sufficient conditions for the exercise of the requisite functions." Suitable external conditions will enable the exercise of the function, she explains. Of course, even internal capabilities will have developed relationally in the first instance, so that relationality remains central in the social conditions for capabilities. The disabled student above might have the internal capability to engage critically, but finds herself excluded from functioning in learning contexts where appropriate learning support is not provided. In Nussbaum's conceptualization, this student has the internal capabilities to handle academic work, but the (external) conditions in the institution to enhance her capabilities are missing or constrained. In the capability approach, education is assumed (and expected) to be empowering and transformative. Sen's approach does not allow that education in schools, colleges, and universities might not always operate as the unqualified good that he takes it to be (Unterhalter 2003a). But education of poor quality is a disadvantage, and one that can persist throughout a lifetime. Our positive and negative experiences of formal education at schools, colleges, and universities will affect choices that we make and how we navigate our futures. Such experiences include curriculum, pedagogy and assessment, and the culture of the school, particularly whether or not all students are equally valued and respected. For example, a study of mature learners in Scotland by Gallacher et al. (2002) documented the continuing negative impact of earlier schooling experiences on people's perceptions of themselves as successful learners. One interviewee said: "I had problems with my level of self esteem connected with my past educational experience. The discouraging thing is really inside me. It is this internal stuff that always comes back and beats me up" (2002, p. 506). A deep sense of failure at school might then reduce the chances of further educational agency and freedom. In Unterhalter's (2003a) account of gender inequality in South African schools, endorsed by George (2004), it is clear that harassment of female students by their teachers and male peers greatly diminishes their capability to succeed at or even to stay in school. Capabilities can be diminished as well as enhanced.

These examples highlight a number of issues concerning an assessment of capabilities. First, given that the process of conversion will draw on intersecting differences and might require trade-offs between equalizing capabilities in one space but not another, how do we distinguish between these fairly? Second, given that in some societies school might provide reasonable social arrangements for some capabilities (for example, gaining literacy or the knowledge to pass a degree), but not others (for example, being free from sexual harassment or the experience of racial discrimination), how can one maximize equality in the capability space? We return to these issues in the Conclusion.

Selecting Capabilities and the Question of a List

Sen and Nussbaum each take a different approach to the matter of selecting capabilities. Sen's approach is grounded in participatory human development, and Nussbaum's in analytic philosophy. So Sen (1992, 1999,

2004a, 2004b, 2002) has consistently argued for the importance of public participation and dialogue in arriving at valued capabilities for each situation and context. His capability approach is deliberately incomplete; he does not seek a complete ordering of nonnegotiable options. He does not stipulate which capabilities should count, nor how different capabilities should be combined into an overall indicator of well-being and quality of life. For him a "workable solution" is possible without complete social unanimity. He argues that all the members of any collective or society "should be able to be active in the decisions regarding what to preserve and what to let go" (1999, p. 242). There is a real social justice need, Sen says, "for people to be able to take part in these social decisions if they so choose" (1999, p. 242). The process of public discussion is crucial, so that the public as much as the individual is seen to be an active participant in change, as citizens whose voices count.

In Sen's model of deliberation, a dialogical democratic process encourages open and public debate, discussion, and dispute around proposals for development or development priorities in order to arrive at a collective and reasoned determination of what are the best policies and capabilities. Those affected by any policy or practice should be the ones to decide on what will count as valuable capabilities. Opportunities "are also influenced by the exercise of people's freedoms, through the liberty to participate in social choice and in the making of public decisions" (Sen 1999, p. 5). Sen is, therefore, quite critical of the idea that "pure theory" can substitute for the "reach of democracy," or that a list of capabilities could be produced irrespective of what the public (or publics) understands and values. He argues strongly that to insist on "a fixed forever list of capabilities would deny the possibility of progress in social understanding"(2004a, p.80). This emphasis on participation in turn, however, makes much of capabilities for literacy and conditions of free and accessible media. Because Sen is concerned as much with the process—who is making the choice—by which we arrive at an outcome as the outcome itself, this must also then mean paying attention to theories of deliberative democracy, theories of power, and theories of voice and participation.

Freedom, for Sen (following educators such as John Dewey and Paolo Freire), is concerned as much with the processes of decision making as with the opportunities to achieve valued outcomes. In other words, we make development and freedom by *doing* development and freedom. Thus a list of capabilities in education or any other area cannot simply be prespecified without public consultation. Freedom is intrinsically important "in making us free to choose something we may or may not actually choose" (Sen 1999, p. 292). Moreover, Sen argues that a space for justice or fairness does exist in the human mind and that "basic ideas of justice are not alien to social beings." At issue is making "systematic, cogent and effective use of the general [moral] concerns that people have" (1999, p. 262).

Nussbaum (2000) has a somewhat different view on the idea of a capabilities list. She argues for a Marxian/Aristotelian conception of truly human functioning as "the proper function of government," which requires that it "make available to each and every member of the community the basic necessary conditions of the capability to choose and live a fully human good life, with respect to each of the major human functions included in that fully good life" (Nussbaum 1993, p. 265). She has produced, and robustly defends (for example, see Nussbaum 2003a) a universal, cross-cultural list of central capabilities for human flourishing and a life of dignity because, she points out, "we need to have some idea of what we are distributing, and we need to agree that these things are good" (Nussbaum 1998, p. 314). Nussbaum's list comprises life; bodily health; bodily integrity; senses, imagination, and thought; emotions; practical reason; affiliation; other species; play; and control over one's environment (2000, pp. 78–80). These ten central human capabilities would need to be present for such a fully human good life (1998, 2003a, 2000). Nussbaum, therefore, seeks to give a specific content to capabilities, arguing that Sen's reluctance to make commitments about what capabilities a society ought centrally to pursue means that guidance in thinking about social justice is too limited. Her list, she claims, constitutes "a minimum account of social justice" (2003a, p. 40). Nussbaum (2000, 2003b, 2004) further argues that the list is humble, open-ended, and revisable, although it is not clear who will revise it.

Nussbaum's list has been criticized (e.g., by Robeyns 2003), insofar as we do not know through what, if any, process of public debate the list was produced, whose voices were heard, or how it is to be revised and by whom. For Sen (2004a) it is "canonical," and this is problematic for him; as noted above, he leaves his framework deliberately vague, because of the importance for him of communities deciding what capabilities count as valuable. Nussbaum refutes the charge (e.g., in Barclay 2003) that her list prescribes a comprehensive notion of the good (Nussbaum 2003b, p. 27) and argues that she supports a position grounded in political liberalism with an overlapping consensus for the content of her list among people "holding a wide range of divergent reasonable comprehensive doctrines" of what is good. While Sen (2004a, p. 79) has said that he is not opposed to lists per se "so long as we understand what we are doing (and in particular that we are getting a list for a particular reason, related to a particular assessment, evaluation or critique)," he would argue that to specify a single list of capabilities is to change the capability *approach* into the capability *theory,* whereas he intends it to operate as a general framework for making normative assessments about the quality of life (Robeyns 2003). In this volume, the problem of lists is addressed by Janet Raynor (chapter 8), Melanie Walker (chapter 9), and Mario Biggeri (chapter 10).

The Specificity of Education and Learning

Over and above the problem of lists, the specificity of education and learning raises some particular issues for the approach: first, the question of children and second, the question of what counts as education and learning. How do we assess the capabilities of children? Should children be allowed to make their own choices to accept or to reject education or certain components of their schooling? Do we need some theoretical understanding of the "education good"? How do we judge who in education is lacking capabilities central to learning in school and for postschool opportunities and choices? Surely if we are going to raise questions about equality in education, we need some idea of what we take to count as "education." As Nussbaum (2003a) robustly states, unless we do this we end up saying that we are for justice but that any old conception of justice is all right with us. Robeyns (2003) points out that while at some abstract philosophical or theoretical level one might argue that all valuable capabilities matter, "this is no option for second-best [nonideal] theorizing or for applications" (p. 35). Application in education requires us to address aspects of the indexing problem and what capabilities matter.

We think it important to give content to our notion of capability in and through education. We thus need to engage the view that not everything counts as education if we wish at one and the same time to argue that education expands human freedoms, agency, and empowerment.

With regard to education and children, Sen emphasizes the importance of schooling to nurture future capabilities (Saito 2003). Nussbaum, too, is clear that children should be required to remain in compulsory education (schooling) until they have developed the capabilities that are important in enabling them to have genuine and valued choices, such as the choice to exit from a traditional, religious community. She concedes that we cannot develop a mature adult capability without having some practice of it. She states that "education in critical thinking and debate is a compelling state interest" and that children taught to develop (learn) these capabilities in debating complex and controversial social and moral issues "can always reject the teaching later" (2003b, p. 42).

This highlights how capabilities cannot be evaluated without any sense of the link with functionings. Nussbaum points out we might need to promote a relevant capability "by requiring the functioning that nourishes it"

(2000, p. 91). She gives the example of requiring children to spend time in play, storytelling, and art activities as a way to promote the general capability of "play," which is important for adults. In other words, they need to function in order to develop the mature capability. Brighouse (2002) argues that it is crucial for children and young adults to practice critical thinking and reflection, and for us to evaluate their functioning in these areas in order that they might develop and enhance this capability through education.

In children and young people's education it therefore makes sense to consider people's functionings (what we manage to achieve) and not just capabilities. Thus, teachers need to know if and how capability is being developed, by whom, and under what conditions, as well as how this relates to capabilities.

It is clear that addressing the problem of children, capabilities, and functionings raises issues about the content of education capabilities. Brighouse and Swift (2003, p. 367) point out that education is not a neutral activity; it always embodies a view about what is good in human life, otherwise it might "seem vapid, even pointless." But are there education capabilities that we might argue are objectively good for an individual's educational development? We might not wish to describe as education a process that tolerates, ignores, or even encourages prejudice, exclusion, marginalization, or harassment of any student on the basis of difference, or that limits their access to knowledge or critical and confident participation in learning. Education that contributes to *un*freedoms would be deeply incompatible with the capability approach. We need to be clear that respecting a plurality of conceptions of the good life (and hence of how education is arranged) is not the same as endorsing all versions of the good life, and this has clear educational implications. The key issue here is that to count as *education,* processes and outcomes ought to enhance freedom, agency, and well-being by "making one's life richer with the opportunity of reflective choice" for a life of "genuine choices with serious options" (Sen 1992, p. 41), and enhancing "the ability of people to help themselves and to influence the world" (Sen 1999, p. 18). The process of identifying education capabilities appears to entail some form of participatory and inclusive dialogue, however conceptualized.

Sen promotes the notion of the capability of the individual agent to critically reflect and make worthwhile life choices from the alternatives available to her. The point is that capability, he would argue, equips us to determine our own major goals in life, and we should not prescribe for adults how they should live. But we still need to sort out the matter of developing capability only or also functioning, for example, reading, writing, and critically assessing information. At issue is that capabilities are counterfactual and difficult to measure or evaluate. In the matter of learning we may need to evaluate functioning as a proxy for capability (Gaspar and Van Staveren 2003; Robeyns 2005). The point is that in education we probably do have to evaluate functioning, but we need to do this without prescribing to students the choices they make about their own lives, and respecting a plurality of conceptions of the good life within a democratic society.

The chapters in the book explore many of these issues raised in the introduction in greater depth. We have organized the book into three parts. Part I consists of chapters that place their primary emphasis on a critical discussion of concepts in the context of some specific conditions in education. The chapters in Part II illuminate how the capability approach can be used to analyze and interpret data concerning education. The distinction between parts I and II is slightly artificial as there is discussion of theoretical issues in the chapters in part II and reflections on data in part I, but the difference is one of emphasis. Readers who wish to start with a "feel" of what the capability approach looks like when applied to education are advised to start with part II and come back to part I for discussions and debates about some of the conceptual issues. Readers who want to immerse themselves in the conceptual debate before looking at analyses of particular contexts should start with part I and then move to part II. To follow up the ideas discussed in this book there is a bibliography on capabilities and education after the concluding chapter. The concluding chapter assesses some of the scholarship to date in this field and highlights a number of difficult areas where debate on the approach is ongoing.

References

Aikman, Sheila. 1999. *Intercultural education and literacy: An ethnographic study of indigenous knowledge and learning in the Peruvian Amazon Studies in Written Language and Literacy 7.* Amsterdam: John Benjamins.
Alkire, Sabina. 2002. *Valuing Freedoms: Sen's capability approach and poverty reduction.* Oxford: Oxford University Press.
Ball, Stephen. 2003. *Class strategies and the education market: The middle classes and social advantage* London: Routledge Falmer.
Barclay, Linda. 2003. What kind of liberal is Martha Nussbaum? *Nordic Journal of Philosophy,* no. 4:5–24.
Bourdieu, Pierre, and J-C Passeron. 1977. *Reproduction in education, society and culture.* 2nd ed. London: Sage.
Bowles, S., and H. Gintis. 1976. *Schooling in capitalist America.* New York: Basic Books.
———. 2002. Schooling in capitalist America revisited. *Sociology of Education* 75 (2): 1–18.

Brighouse, Harry. 2000. *School choice and social justice*. Oxford: Oxford University Press.

———. 2002. What rights (if any) do children have? In *The moral and political status of children*, edited by A. Archard and C. Macleod. Oxford: Oxford University Press.

Brighouse, Harry, and Adam Swift. 2003. Defending liberalism in education theory. *Journal of Education Policy* 18:355–373.

Comim, Flavio. 2003. Capability dynamics: The importance of time to capability assessments. Paper read at the Third International Conference on the Capability Approach, September, University of Pavia, Italy.

Dreze, Jean, and Amartya Sen. 1995. *India: Economic development and social opportunity*. Oxford: Oxford University Press.

Gallacher, Jim, Beth Crossan, John Field, and B. Merrill. 2002. Learning careers and the social space: Exploring the fragile identities of adult returners in the new further education. *International Journal of Lifelong Education* 21:493–509.

Gaspar, Des, and Irene Van Staveren. 2003. Development as freedom—and as what else? *Feminist Economics* 9:137–161.

George, Erika R. 2004. Human rights, development and the politics of gender based violence in schools: Enhancing girls' education and the capabilities approach. Paper read at the Fourth International Conference on the Capability Approach: Enhancing human security, September 5–7, University of Pavia, Italy.

Kwesiga, J. 2002. *Women's access to higher education in Africa: Uganda's experience*. Kampala: Fountain Publishers.

Lynch, Kathleen, and John Baker. 2005. Equality in education: An equality of condition perspective. *Theory and Research in Education* 3:131–164.

McLeod, Julie. 2005. Feminists re-reading Bourdieu: Old debates and new questions about gender habitus and gender change *Theory and Research in Education* 3:7–9.

Nussbaum, Martha C. 1993. Non-relative virtues: An Aristotelian approach. In Nussbaum and Sen, *The quality of life*.

———. 1997. *Cultivating humanity. A classical defence of reform in liberal education*. Cambridge, MA: Harvard University Press.

———. 1998. The good as discipline, the good as freedom. In *Ethics of Consumption*, edited by D. A. Crocker and T. Linden. Lanham: Rowman and Littlefield.

———. 2000. *Women and human development: The capabilities approach*. Cambridge: Cambridge University Press.

———. 2002. Education for citizenship in an era of global connection. *Studies in Philosophy and Education* 21:289–303.

———. 2003a. Capabilities as fundamental entitlements: Sen and social justice. *Feminist Economics* 9 (2–3): 33–59.

———. 2003b. Political liberalism and respect: A response to Linda Barclay. *Nordic Journal of Philosophy* 4:25–44.

———. 2004. Liberal education and global community. *Liberal Education* (Winter).

———. 2006. Education and democratic citizenship: Capabilities and quality education. *Journal of Human Development* 7 (3): 385–398.

Nussbaum, Martha C., and Amartya Sen, eds. 1993. *The quality of life: Studies in development economics.* Oxford: Oxford University Press.

Robeyns, Ingrid. 2003. Is Nancy Fraser's critique of theories of distributive justice justified? *Constellations* 10:538–553.

———. 2005. The Capability Approach: A theoretical survey. *Journal of Human Development* 6 (1): 93–114.

Saito, Madoka. 2003. Amartya Sen's capability approach to education: A critical exploration. *Journal of Philosophy of Education* 37 (1): 17–33.

Sen, Amartya. 1980. Equality of what? In *The Tanner Lectures on Human Values,* edited by S. McMurrin. Salt Lake City: University of Utah Press.

———. 1992. *Inequality re-examined.* Oxford: Oxford University Press.

———. 1993. Capability and well being. In Nussbaum and Sen, *The quality of life.*

———. 1999. *Development as freedom.* Oxford: Oxford University Press.

———. 2002. *Rationality and freedom.* Cambridge, MA: Harvard University Press.

———. 2004a. Capabilities, lists and public reason: Continuing the conversation. *Feminist Economics* 10:77–80.

———. 2004b. What Indians taught China. *The New York Review of Books* 19:61–66.

Stromquist, Nelly. 1998. Empowering women through knowledge: Politics and practices in international cooperation in basic education. Stanford, CA: SIDEC.

Terzi, Lorella. 2005. Beyond the dilemma of difference: The capability approach to disability and special educational needs. *Journal of Philosophy of Education* 39:443–460.

Unterhalter, Elaine. 2003a. The capabilities approach and gendered education: An examination of South African complexities. *Theory and Research in Education* 1 (1): 7–22.

———. 2003b. Crossing disciplinary boundaries: The potential of Sen's capability approach for sociologists of education (review). *British Journal of Sociology of Education* 24 (5): 665–669.

Part I

Theoretical Perspectives on the Capability Approach and Education

Editors' Introduction

The chapters in this part set out to examine some of the concepts the capability approach draws on and to consider its explanatory reach in relation to education considered as a field of policy, practice, and social relations. Themes of justice, freedom, and equality recur throughout these chapters. A number of authors consider some of the critiques of the capability approach and how salient problems of open-endedness are. Two chapters (Unterhalter and Vaughan) are particularly concerned with gender and consider the light the capability approach sheds on different dimensions of gender inequalities.

The writers approach these issues from different disciplinary backgrounds and this has a bearing on the features they emphasize. Terzi and Brighouse are philosophers, Flores-Crespo is an economist, and Unterhalter and Vaughan work in the field of international and comparative education. Though all are concerned with conceptual problems, some discussions are more abstract, while others consider particular issues of how the capability approach can aid understanding of issues relating to education, global justice, or gender.

The chapters single out different areas of education for discussion. Terzi is concerned with the capability to be educated and therefore with basic levels of education for children and adults. Flores-Crespo seeks to distinguish the capability approach from human capital theory and is interested in how policy and pedagogy that take their bearings from the capability approach suggest different relations between education and political economy. He uses a number of examples from higher education. Unterhalter and Brighouse do not specify a particular level of education as their main focus of concern, but in seeking to disentangle the global aspiration for Education for All (EFA) from the meaning it has acquired denoting minimal provision, they tend to highlight examples concerned with elementary education. Vaughan, too, is concerned with this level. Unterhalter's discussion of meanings of gender equality in education ranges widely across different phases and is concerned with teacher training, secondary schooling, as well as relations in schools.

Although these chapters are primarily conceptual rather than empirical, they draw on examples from diverse settings. Terzi considers empirical work on education from the UK and the United States, Flores-Crespo from the United States and Peru, Brighouse and Unterhalter in their coauthored piece from the reviews made by multilateral organizations such as UNICEF and

UNESCO, and Unterhalter in her single-authored piece on gender equality draws on problems posed in the reform of education in Kenya. Vaughan uses examples from the UK and India. The capability approach thus appears to have resonance in a wide range of educational settings.

In opening this part, Lorella Terzi explores how the capability to be educated might be conceptualized, showing how the absence or lack of education would harm an individual. Education plays a substantial role in the expansion of other capabilities at any particular moment and affects the potential of future capabilities. The capability to be educated is fundamental and foundational to different capabilities to lead a good life. She then turns to developing a methodology for the functionings and capabilities constitutive of education and concludes with a robust analysis of why the provision of education is a matter of justice.

Pedro Flores-Crespo draws out limitations in how education is understood in human capital theory, still a major paradigm in thinking about the purposes and outcomes of education. He contrasts this with other ideas of education concerned with human agency, developed over many centuries, which are more in tune with the concern with freedom in the capability approach. He then turns to some of the issues that arise owing to the institutional location of education and considers how problems of policy and pedagogy bear on how freedom is centrally positioned in thinking about education and capabilities.

Elaine Unterhalter and Harry Brighouse discuss the global social justice movement to establish EFA and question the approaches to measurement that have been used by UN agencies. In trying to develop an alternative approach to measurement that draws on the capability approach, they seek to defend the approach from some of its critics, notably Thomas Pogge, who considers that capabilities are not superior to resources when thinking about justice. They defend an approach to measurement that attempts to capture instrumental, intrinsic, and positional dimensions of education while emphasizing the centrality of freedoms.

Elaine Unterhalter's second piece explores the question of what gender equality in education means and how the capability approach has affinities with particular meanings of gender and equality. She shows how the approach to education equality in the capability approach is different, on the one hand, to that which emphasizes equality in the distribution of inputs or outputs of education, and on the other, from that which stresses empowerment and equal conditions in education. In this the chapter has relevance for thinking of other inequalities in education, for example, those relating to class, race, or ethnicity.

Rosie Vaughan considers gender and outcomes from education. She highlights some of the shortcomings in existing approaches to measurement and then considers how understanding processes that bear on the capability set allow us to better understand how education works both as an important capability in its own right and as a capability that facilitates the development of other capabilities. She also examines Sen's distinction between well-being and agency in relation to capabilities to participate in education and develop capabilities through education.

While the chapters in this part of the book deal primarily with abstract ideas about justice, equality, freedom, and ethics, they talk to very concrete situations teachers, parents and education officials grapple with daily. The capability approach, as an idea in social justice, needs to be assessed in relation to these everyday concerns of policy and practice that will make education transformative.

2

The Capability to Be Educated

Lorella Terzi

In this chapter I outline a possible conception of the capability to be educated. I argue that the capability to be educated, broadly understood in terms of real opportunities both for informal learning and for formal schooling, can be considered a basic capability in two ways. First, in that the absence or lack of this opportunity would essentially harm and disadvantage the individual. Second, since the capability to be educated plays a substantial role in the expansion of other capabilities, as well as future ones, it can be considered basic for the further reason that it is fundamental and foundational to the capabilities necessary to well-being, and hence to lead a good life. Finally, I argue that this conception highlights how the capability to be educated constitutes a fundamental entitlement, and why its provision becomes a matter of justice. The key issue taken up is that the capability approach requires focusing on the contribution that the capability to be educated makes to the formation and expansion of human capabilities, and hence to the contribution it makes to people's opportunities for leading flourishing lives.

Amartya Sen identifies basic capabilities as a subset of all capabilities. Basic capabilities, in his approach, are "a relatively small number" of centrally important beings and doings that are crucial to well-being (Sen 1992, p. 44). The capabilities to be well-nourished and well-sheltered, to escape avoidable morbidity and premature mortality, to be educated and in good health, and to be able to participate in social interactions without shame, are all examples of basic capabilities.[1]

The capability to be educated is included among these fundamental capabilities. In his analysis of development and poverty, Sen highlights the contribution of education to the quality of life and the formation and expansion of human capabilities. However, despite this important role, in

Sen's approach education is generically referred to as basic, elementary education, and mainly expressed in terms of levels of literacy. Hence, the conceptual and normative implications of the basic capability to be educated remain unspecified.

I turn now to addressing some of these implications.

Understanding Basic Capabilities

The question of determining basic capabilities relates to the possibility "that some capabilities may be so basic to human welfare that they can be identified without any prior knowledge of the particular commitments that are held and expressed by an individual or group" (Alkire 2002, p. 154). Sen has addressed issues of basic capabilities in his analysis of poverty. He maintains that, rather than in terms of income inadequacy and relatively to the position enjoyed by others in society, poverty is best addressed in terms of "basic capability failure," namely the absolute inability of individuals and communities to choose some valuable beings and doings that are basic to human life (Sen 1992, p. 109). Basic capabilities are therefore a subset of all capabilities and refer to the possibility of satisfying "certain crucially important functionings up to certain minimally adequate levels" (Sen 1980, p. 41; 1992, p. 45; Robeyns 2001, p. 11). Sen does not provide a definite list of basic capabilities, nor a fully justified account of how to identify them, but he mentions several elementary capabilities, which include the capability to be sheltered, nourished, educated, and clothed (Sen 1999, p. 20; 1992, p. 69). He furthermore specifies that, given the "ambiguity of the concept of basicness" (1992, p. 45), the term "basic capability" is open to different interpretations.

This unspecified dimension is reflected in the different conceptions of the idea of basic capability, both within the capability approach and in its empirical applications. In what follows I shall briefly analyze two important specifications of the idea: first, Sabina Alkire's (2002) operationalization of basic capabilities as capabilities to meet basic needs and, second, Bernard Williams's (1985) understanding of basic as fundamental capabilities. I maintain that these perspectives provide interesting theoretical frameworks, which can be fruitfully applied to the conception of the capability to be educated. Let us now consider them.

Sabina Alkire's work (2002) on the operationalization of the capability for poverty reduction presents an understanding of basic capabilities in relation to the idea of human needs. Her account draws, on the one hand, on the literature on human needs, and on the other, develops "a conception of basic human needs that closely relates to Sen's work. It defines basic needs with reference to absolute harm, rather than to wants, needs, desires, or preferences"

(2002, p. 157). In Alkire's view basic needs are described, first, with reference to the substantive functioning that is harmed if the basic need is unmet. For example, not meeting the functionings of being well nourished or being clothed fundamentally harms the individual. Second, basic needs are expressed at a sufficient level of generality, in that they refer to "what is needed at a general level," for instance, and following Sen's own indication, shelter, nutrition, education, and clothes (Alkire 2002, p. 160). Therefore, expressing the basic need to be nourished at a general level implies reference to dietary requirements rather than to specific accounts of the food to be provided. Alkire's two criteria for identifying basic capabilities relate to what is fundamental in order to avoid harm, and, to a level of generality, which allows basic capabilities to be applied to different situations (2002, p. 160). A further specification of these basic capabilities is required only at the level of their operationalization in different contexts, cultures, and societies.

Based on these criteria, Alkire outlines the following conception of basic capability, which includes a specific understanding of basic needs but inscribes it in the framework of capability:

> A basic capability is a capability to enjoy a functioning that is defined at a general level and refers to a basic need, in other words a *capability to meet a basic need* (a capability to avoid malnourishment; a capability to be educated, and so on). The set of basic capabilities might be thought of as capabilities to meet basic human needs.
>
> (2002, p. 163)

Notice here that Alkire's conception of basic capability retains the strong sense of needs as one's fundamental requirements while, at the same time, grounding it in the important concept of potential for intentional choice implied in the idea of capability. This allows for people's deliberate choice to refrain from meeting certain basic needs in order to pursue other aims, providing that the relevant capabilities of meeting basic needs are still retained. As she illustrates,

> a hunger striker or a Brahmin may regularly refrain from eating, because they personally value the religious discipline or the exercise of justice-seeking agency, but the side effect of pursuing these is that they will not be well nourished . . . while the Brahmin's "functioning" of being well-fed would indeed be blighted by fasting, her *life* might be regal and radiant.
>
> (2002, p. 171)

Thus, what Alkire brings to the fore is the fundamental element of choice, constitutive of and explicit in the concept of capability, and its relation to the pursuit of people's valuable ends and objectives, and hence of their well-being. Both are fundamental dimensions that the capability approach explicitly provides with respect to accounts based on basic human needs.

Alkire maintains, furthermore, that another crucially important element highlighted by the capability approach is to make explicit the fundamental dimension of participation. She illustrates this point by providing the example of two countries, A and B, whose goal is identified in terms of meeting basic needs such as nourishment, shelter, education, and health. If country A had better results in meeting these basic needs than country B, we would say that A is better than B. And we would have to reach this conclusion even when A had achieved its increase by means of coercion, for instance, by the government. To evaluate this situation differently, we should reframe the initial aims to include among the basic needs also elements of choice, participation, and freedom, all elements fundamentally implied by the concept of capability (Alkire 2002, p. 170). Consequently, the important insight of the capability approach consists exactly in this explicit and crucial focus on people's choice and participation related to the element of freedom, which is constitutive of the concept of capability. These considerations allow Alkire to conclude that the capability approach, when compared to other perspectives, and specifically to human needs theories, and in operational terms, "is a wider, philosophically more rigorous way of conceiving poverty reduction in relation to the full life" (2002, p. 167). This is owing to the explicit and consistent value it assigns to people's choice and participation and their relation to freedom in the pursuit of well-being (2002, p. 170).

Ultimately, analyzing Sen's concept of basic capability and its specification by Alkire has highlighted how the idea of basic capability, understood as the capability to meet basic needs, inscribes it in a substantial philosophical framework. Fundamental elements of this perspective are the overall aims of well-being and individual's choice and participation. If Alkire's account of basic capabilities presents potentially interesting implications for understanding education as basic capability, Williams's view on basic capability seems also particularly relevant in thinking of education.

Williams understands basic as fundamental capabilities in the sense of some invariant, underlying capabilities that are "derived from some universal and fundamental facts about human beings" (Williams 1985, p. 101). His reference to the capability to appear in public without shame is particularly useful in understanding the precise meaning of fundamental capabilities. Williams recalls Sen's use of Adam Smith's example of the man who cannot appear in public without shame, given the cultural and social arrangements he lives in, unless he can wear a linen shirt. In Williams's understanding, although the requirements in order to appear in public without shame depend on specific contexts, the invariant, fundamental capability at play here is the capability to "command the material of self-respect" (1985, p. 101). And it is in this sense, according to Williams, that certain capabilities are basic, in that they are fundamental to human well-being. Moreover, these capabilities are distinct from more trivial ones, such as those associated with the commodities of choosing from such an increased range of washing powder (1985, p. 98).

According to Robeyns, basic capabilities in Williams's specification have to be understood as "deeper, foundational, generic, general, aggregated (not over persons but over different capabilities in one person) capabilities" (Robeyns 2001, p. 12). Robeyns notices that the use of basic in the sense of fundamental capabilities is implemented in several empirical studies based on the capability approach. These refer to fundamental capabilities including the capability of being sheltered and living in a safe environment; being in good health and enjoying physiological well-being; being educated and having knowledge; enjoying social relations and emotional well-being; as well as being safe and maintaining bodily integrity.

How do these fundamental capabilities compare with those outlined as basic in the sense of meeting basic needs? Here Robeyns's suggestion appears particularly interesting. She maintains that

> a person's capabilities consists of a number of fundamental capabilities which are each made up by a number of more specific capabilities, some of which are basic and some of which are non-basic. The basic capability of a person is then some kind of aggregate of the basic capabilities in each of these different fundamental capabilities.
>
> (Robeyns 2001, p. 13)

According to this conception, we can think of the fundamental capability of being sheltered and living in a pleasant and safe environment as including basic capabilities such as having a place to live, living in a safe area, and having access to water, and nonbasic capabilities such as choosing one's water supplier. Hence fundamental capabilities are broader than basic capabilities, but retain the foundational importance for well-being expressed by basic capabilities.

Let us now summarize the main points of the discussion. First, basic capabilities can be conceptualized as capabilities to meet basic needs. Here the idea of basic capability retains the fundamental requirement associated with human needs, while inscribing it in a philosophical approach concerned with people's freedoms and well-being. Second, there is a fundamental and foundational dimension inscribed in the idea of basic capability.

Both these views provide a framework for conceptualizing the basic and fundamental capability to be educated as essential to the expansion of future capabilities, and therefore as constitutive of an entitlement in education. The next section will explore this conception.

The Capability to be Educated as a Basic Capability

The capability to be educated can be considered basic in two interrelated respects. First, in that absence or lack of opportunities to be educated would essentially harm or substantially disadvantage the individual. Education thus conceived responds to the basic need of the individual to be educated. Second, since the capability to be educated plays a substantial role in the expansion of other capabilities as well as future ones, it can be considered fundamental and foundational to different capabilities, and hence inherent to the very possibility of leading a good life. Following this conception, I maintain that the capability to be educated entails the selection of specific subsets of enabling conditions that are fundamental to it. It also requires a set of methodological criteria for the selection. Let us analyze this understanding in more detail.

The first facet in which the opportunity to be educated can be considered a basic capability relates to its crucial importance for people's well-being. The capability to be educated is basic, since absence or lack of education would essentially harm and disadvantage the individual. This is specifically, albeit not solely, the case for children, where absence of education, both in terms of informal learning and schooling, determines a disadvantage that proves difficult, and, in some cases, impossible to compensate in later life. Perhaps the most striking example of this need for education is represented by the case of feral children. Studies of feral children,[2] children who lived in the wild or in cages, and deprived of any form of learning for a substantial part of their childhood, show the profound harm caused by the absence of education. In these cases, not only are language functionings and broader communicative functionings substantially harmed, but reasoning and learning functionings are also compromised. This highlights the importance of education for the formation of human capabilities and, more generally, appears to confirm the understanding of the capability to be educated as responding to a person's basic need, in its specification in terms of avoiding harm and disadvantage.

However, a further aspect of the capability to be educated as essential requirement relates to its greater context-dependence if compared, for instance, to the capability to be well nourished. It seems that the capability to be educated as essential in order to avoid disadvantage to the individual implies considerations about the design of social arrangements, which are more relevant in the case of education than in that of hunger. Hence, determining the level at which a person is considered well nourished seems at least more straightforward than adjudicating the level at which a person is educated. This relates to considerations about the complexity of education, which are well captured in the second understanding of the basic capability to be educated.

The capability to be educated is basic also in the sense of being a fundamental capability, and foundational to other capabilities as well as future ones. Consider, for instance, the case of learning mathematics. Formally learning mathematics not only expands the individual's various functionings related to mathematical reasoning and problem solving, but also widens the individual's sets of opportunities and capabilities with respect to, on the one hand, more complex capabilities and, on the other, better prospects for opportunities in life. The broadening of capabilities entailed by education extends to the advancement of complex capabilities, since while promoting reflection, understanding, information, and awareness of one's capabilities, education promotes at the same time the possibility to formulate exactly the valued beings and doings that the individual has reason to value (Saito 2003). On the other hand, the expansion of capabilities entailed by education extends to choices of occupations and certain levels of social and political participation. Learning mathematics may lead to choosing to become an economist or a mathematics teacher, for instance, as well as promote one's civic participation in different forms. These considerations lead to an understanding of the capability to be educated as a fundamental capability, which includes basic capabilities in terms of those enabling beings and doings that are fundamental in meeting the basic need to be educated, and are equally foundational to the promotion and expansion of higher, more complex capabilities.

Thinking of education in the above meaning relates substantially to an understanding of education as a complex good entailing instrumental and intrinsic values (Brighouse 2000; Saito 2003; Swift 2003; Unterhalter and Brighouse 2003). Education has an instrumental aspect, since it is a means to other valuable goods, such as better life prospects, career opportunities, and civic participation. It improves one's opportunities in life. In this sense education, and specifically schooling, promotes the achievement of important levels of knowledge and skills acquisition, which play a vital role in agency and well-being. On the other hand, education is intrinsically good, is valuable in itself, in that being educated, other things being equal, enhances the possibility of appreciating and engaging in a wide range of activities that are fulfilling for their own sake. For instance, being initiated through education into the appreciation of poetry, or aspects of the wildlife in natural environments, or different kinds of music, relates to a personal fulfillment that is not instrumental in securing better jobs or positions, but brings about a more fulfilling life. Ultimately, the instrumental and intrinsic aspects of education relate to the enhancement of freedom, both in terms of well-being freedom and agency freedom, which are aspects central to and highlighted by the capability approach. The fundamental contribution of education to the flourishing of individuals and their quality of life is well attested by empirical research. Sen highlights, for instance, the benefits related to the education of women, both in broadening their freedom to exercise agency and in its correlation to a reduction in infant mortality (1999, p. 198). Nussbaum (2003) defends the value of education as crucial not only to human dignity, but also and specifically to the promotion of women's capabilities in many areas of their lives. Furthermore, numerous studies show significant statistical correlations between education and changes in different aspects of people's lives: better educated people live longer, healthier lives and transmit more material as cultural benefits to their children (Ferri, Bynner, and Wandsworth 2003). A recent qualitative study in the UK on the wider benefits of learning (Schuller et al. 2004) shows that education has an impact on people's psychological and physical well-being, family life, and communication between generations, as well as people's ability and motivation to take part in civic life. The study furthermore highlights the benefits of education both for agency freedom and well-being freedom. As Schuller et al. write:

> The important function it [education] provides is enabling people to have a sense of a future for themselves, for their families and perhaps also for their communities, which they can to some extent control or influence. Several of our respondents spoke about having a sense of agency which they did not have before. In other words, education provides a kind of choice in life . . . the notion of choice (and therefore some degrees of personal autonomy) is present in ways which did not previously exist and horizons are extended beyond what might have been imagined.
>
> (2004, p. 190)

These empirical studies confirm the value of the capability to be educated as fundamental as well as foundational to different capabilities. They show the role of education as having sustained and transformative effects, as well as highlighting the interconnectedness of the effects of learning in promoting people's well-being and therefore their freedom to lead flourishing lives. Ultimately, the capability to be educated can be considered basic both in the sense of being essential to well-being—thus avoiding harm—and related to the previous reason, as it is foundational to the expansion of other, more complex, capabilities.

Having addressed, albeit only very provisionally, the ways in which the capability to be educated can be considered basic, my task is now to outline what functionings and capabilities are constitutive of education thus conceptualized. This task has two interrelated dimensions. The first consists in determining the subset of functionings and capabilities basically constitutive of education, whereas the second refers to the criteria for determining these constituents. It is, in short, the problem of providing a possible list of basic functionings and capabilities in education and determining the principles underlying it. This is a debated problem both "within" and "outside" the capability approach.[3] As it is known, on the one hand, Sen has not provided a definite list of valuable capabilities and maintains that such a list should be the result of a democratic process involving debate and participation by those who will be affected by the choice. In this sense, the list would be context-dependent. On the other hand, however, Sen has also recognized how basic capabilities imply an absolute level, which is not related to the specification of the context and which can, therefore, be identified independently of the relative picture. Further conceptions within the capability approach have outlined how Sen's framework, as a general normative one, is not in conflict with specifying a list of capabilities aimed at a determined purpose (Robeyns 2003, p. 15). My task is trying to ascertain what functionings and capabilities are constitutive of the basic capability to be educated, and hence a priori of determined contexts, whilst also aiming at operationalizing capability for the determined and specific purpose of education. In this sense, and following from the previous points, the selection of a specific list of educational capabilities seems not only plausible, but also justified. Finally, this task highlights how the criteria for selecting relevant functionings and capabilities play a fundamental role. I shall start by analyzing the criteria.

Recall here Alkire's two main principles in outlining basic capabilities for poverty reduction: capabilities should be identified in terms of capabilities to meet basic needs, and hence avoiding harm to the person, and they should be expressed at a general level. I maintain that these two principles by which the capability to be educated is selected as a basic capability are workable also at the level of identifying the subset of capabilities constitutive of it. If being educated is basic in terms of being fundamental to well-being, then its components are equally fundamental to it, since they all contribute to avoid harm or disadvantage, thus meeting the first criterion. Furthermore, they can be expressed at the requested level of generality, thus meeting the second criterion (an aspect that I shall address in more detail below).

However, applying these two criteria at the level of identification of functionings and capabilities constitutive of a capability—that of being educated— that is already expressed as basic implies explicitly addressing a potential theoretical problem. This consists in avoiding the possibility of an infinite regress[4] to basic and yet more basic components. Say, we think of being educated as a basic capability, and subsequently specify among its fundamental components the functioning of thinking, and we then proceed to define thinking as a functioning that depends on a more basic functioning, that of wanting to think, and so on; we are caught in a conceptual infinite regress. We need to make sure that the functioning specified is basic and does not imply components more basic to it. Here the identified criteria for selection are crucial in that they have to determine specifically those functionings and capabilities that are absolutely constitutive of education. It is in this sense that the two criteria chosen by Alkire are necessary and applicable to my task, yet perhaps not entirely sufficient to it. In my view, in order to avoid the potential danger of "infinite regress," the criteria have to explicitly include the principle of exhaustion and nonreducibility.[5] The latter criterion requires the elements of the list to be comprehensive, thus including all the important ones, and not overlapping. True, the criterion of avoiding harm could necessarily and sufficiently select only those elements that are basically constitutive of education. Yet, given the complex dimension of education, it seems that a principle explicitly eliciting elements that are exhaustive and mutually exclusive, and hence elements that include all the important and relevant components and that are nonreducible to others, can more effectively select basic capabilities in education.

In her account of capabilities, Robeyns lists five principles upon which to select functionings to address gender inequality (2003). They are: explicit formulation, methodological justification, sensitivity to context, level of generality, as well as exhaustion and non-reduction, which I mentioned above. The first criterion expresses the necessity to formulate the list explicitly and to defend it theoretically. The second refers to expressing the method that has guided the selection for the list, and the third implies acknowledgment and engagement with the debate to which the list itself is relevant (Robeyns, 2003, pp. 15–18). In theorizing basic capabilities in education, we should therefore formulate a list that is explicit and defended also with reference to the methods adopted in devising it and to the relevant educational debate. The first three principles have been indirectly addressed in considering issues relating to the capability to be educated as basic capability and, to a certain extent, can be thought of as subsumed in that. However, the fourth, identifying the appropriate level of generality, which is in common with Alkire's position, is specifically important in determining the basic constituents of the capability to be educated. Robeyns suggests that we can think of the generality of the list at two levels: an ideal level, drawn on procedural aspects, and a more empirical level, where considerations derived from sets of data can modify or alter the original ideal selection (Robeyns 2003, p. 17). In this sense, a list of subsets of capabilities basically foundational to education can be devised at an ideal level, based on principles such as that of avoiding harm and disadvantage to the individual and with comparisons to broad curricular requirements. The same list, however, could be devised at a more empirical level, taking into account the specificity of a certain situation and the availability of sets of data upon which to formulate the selection. Given my aim of theoretically determining a subset of capabilities fundamental to the basic capability to be educated, a certain ideal, general level of specification seems appropriate.

To sum up, the criteria for identifying basic functionings and capabilities in education include the following:

- Functionings and capabilities should be identified in terms of meeting basic needs, and hence avoiding harm and disadvantage
- They should be identified at an ideal level of generality
- They should be exhaustive and nonreducible

These criteria for selecting relevant functionings and capabilities in education provide a methodological basis upon which to proceed to the core of the task at hand: determining what subsets of enabling conditions—beings and doings—are fundamental to the capability to be educated. This is what I shall address in the final section of this chapter.

Selecting Capabilities in Education

What subset of enabling conditions is fundamental to the capability to be educated? Selecting basic capabilities in education is a complex task, and hence my account of this aspect aims necessarily at indicating some possible developments, rather than at providing a complete and exhaustive account. Moreover, selecting basic capabilities in education means looking at what beings and doings are at the same time crucial to meeting basic needs, thus avoiding harm for the individual, and foundational to the enhancement of other beings and doings both in education and for other capabilities. We are looking here at certain enabling conditions whose absence would put the individual at a considerable disadvantage. At the same time, moreover, we are looking at enabling conditions whose exercise is particularly, albeit not solely, important in childhood, since, as Nussbaum notices, "exercising a functioning in childhood is frequently necessary to produce a mature adult capability" (2000, p. 90). Interesting insights in this sense can be drawn on the concept of "serving competencies" and the relative list developed by Charles Bailey (1984) in his analysis of the aims and contents of liberal education.[6] Bailey suggests that a considerable part of education, and especially much of the elementary one,

> must of necessity be instrumental, not in the sense of serving specifically prescribed purposes beyond . . . education, but rather in the sense of making the more substantive objectives of such an education attainable. This is especially the case in the early stages . . . when young pupils must learn how to learn, by acquiring the appropriate means, skills and dispositions.
>
> (1984, p. 111)

Bailey identifies these means, skills, and dispositions as the "serving competencies" needed to achieve subsequent educational aims, and claims that there is indeed little dispute over their choice and relevance. I maintain that the concept of serving competencies presents similarities with that of basic capabilities in education and that the list can provide a useful reference for the selection of basic capabilities. In this sense, considering serving competencies as functional capacities, which allow the attainments of subsequent objectives, implies that their absence would substantially disadvantage the individual. The similarity between the two concepts resides here in the fact that both relate to constitutive elements necessary to the achievement of further aims in education and beyond it. However, the specific instrumental value that Bailey ascribes to serving competencies cannot be referred to basic functionings and capabilities in education. It seems plausible to argue that, given the intrinsic and instrumental value of being educated discussed above, beings and doings fundamental to it are both intrinsically and instrumentally valuable as well. Moreover, this claim appears substantiated when analyzing Bailey's list of serving competencies.

In Bailey's account, serving competencies include literacy, numeracy, logical reasoning, appropriate dispositions, physical fitness, and computer skills. The list, contextualized within liberal education and drawn up when computer technology was still at an initial stage, includes basic elements foundational to education, whose value, however, seems to go beyond the purely instrumental one that Bailey assigns them. Literacy, to mention but one, while being a basic constituent of being educated and while instrumentally allowing the achievement of other educational aims beyond reading and writing as techniques per se, certainly has an intrinsic value as well. Hence serving competencies, understood beyond the level of learning techniques, present an instrumental and indeed an intrinsic value, in that showing a further similarity with basic capabilities in education. However, the interesting insight provided by serving competencies resides primarily in the outlining of some foundational elements, and in their selection in a list. The latter constitutes a useful element of reference and comparison upon which we can build an account of basic enabling conditions.

What are, ultimately, these enabling conditions constitutive of the basic capability to be educated? I suggest the following list of basic capabilities for educational functionings, at the ideal level:

- *Literacy:* being able to read and to write, to use language, and discursive reasoning functionings
- *Numeracy:* being able to count, to measure, to solve mathematical questions, and to use logical reasoning functionings
- *Sociality and participation:* being able to establish positive relationships with others and to participate in social activities without shame
- *Learning dispositions:* being able to concentrate, to pursue interests, to accomplish tasks, to enquire
- *Physical activities:* being able to exercise and being able to engage in sports activities
- *Science and technology:* being able to understand natural phenomena, being knowledgeable on technology, and being able to use technological tools
- *Practical reason:* being able to relate means and ends and being able to critically reflect on one's and others' actions

While presenting relevant similarities with Bailey's serving competencies, this subset of basic capabilities in education complies with the principles outlined as important to its selection, in that the absence of these elements would disadvantage the individual. Moreover, none of the capabilities appears essentially reducible to others, and the list is fairly exhaustive with respect to the foundational elements relevant to education. Furthermore, the list is expressed at a certain level of generality, and hence it allows for more specific lists to be drawn from it in relation to different contexts, cultures, and societies. Finally, the use of "being able to" in expressing capabilities implies here also the opportunity and possibility entailed by the concept of capability, rather than simply the common understanding of "to be able to" in terms of ability. A more detailed analysis of each capability can help in better substantiating this position.

There should indeed be little dispute about literacy as fundamental in education. Listening, speaking, reading, and writing are all essential functionings as well as constitutive of communication functionings and entailing discursive reasoning at different levels. Furthermore, being able to express oneself in different forms, with respect to thoughts as well as imagination, creativity, and belief, is also constitutive of literacy broadly conceived. In this sense, as Bailey notices, "here is the first great practice of human agents into which children must be initiated" (1984, p. 111). Nussbaum (2003) further supports this view in defending the intrinsic value of literacy as crucial to human dignity. One of the most common critiques of literacy relates to its "parochial" and highly Westernized value, one that is allegedly not important to all those individuals who happen to live in nonliterate societies or indeed in communities where being literate is not deemed important. However, as Nussbaum rightly highlights, literacy expands human capabilities and proves to be crucial in progress being made in all areas of people's lives, and especially in the lives of those people, for instance, women, who have been historically and intentionally excluded and prevented from achieving the functioning of being literate. Furthermore, numerous studies attest to the higher level of well-being enjoyed by literate societies when compared to nonliterate ones. Ultimately, all these elements support the importance of literacy in the formation and foundation not only of education capabilities, but also of human ones broadly conceived.

Numeracy, also, pertains to the core of education, and with it functionings such as counting, ordering, comparing, estimating, measuring, and all the functionings related to logical reasoning as one of the ways of making sense of the world and of one's agency in it. Sociality and participation are fundamental functionings in education in different, but related ways. Establishing positive relationships with others allows for personal and social development, which is consistently proven by educators as fundamental to learning. Much learning is promoted and sustained by social functionings such as cooperating, being part of a group, supporting, or being supported by others. Related to sociality, participation is also crucial in education and more so when considering the essential role it plays in the exercise of agency. In this sense, the capability of positively participating in educational activities may well promote the adult mature capability so important for Sen's approach. Learning dispositions entail functionings related to the actual learning process, thus possibilities to concentrate, accomplish tasks, and achieve aims, as well as enquiring and imagining. Physical activities play the important role of maintaining health and general bodily well-being, while also developing bodily awareness and mobility. Science and technology apply to all those possibilities to engage in the understanding of the natural world and its manifestation, as well as developing functionings related to the knowledge and use of technology. Finally, practical reason. Analyzing what constitutes practical reason and its role in education would take this discussion too far from its topical point. However, some considerations may help in justifying its inclusion in this subset of enabling conditions. Nussbaum suggests a notion of practical reason in terms of "the ability to form a conception of the good and to engage in critical reflection about the planning of one's life" (2000, p. 97). She furthermore assigns practical reason a central and crucial role among capabilities, maintaining that it is this kind of reason that makes a life truly human. Although I endorse Nussbaum's position on the importance of practical reason, hers is a substantial notion, whose promotion through education would entail complex and high levels of capabilities. It appears, therefore, that in selecting basic constituents of education, a "thinner" and simpler understanding of practical reason may comply more with the task, while still retaining its crucial importance in enhancing freedom. Hence practical reason in this context is specified as the ability to relate means and ends and to reflect on actions. This, on the one hand, relates to the ability to evaluate and to form independent judgments, while, on the other, establishing the prerequisites for the more mature capability to exercise practical reason in terms of forming a conception of the good and planning one's life.

Each list of specific capabilities is deemed to raise critiques as well as further specifications and amendments. In what follows, I shall try to anticipate some of these by addressing some critical points. In particular, I shall address three sets of considerations: a comparison with Bailey's proposal, a further one with Nussbaum's list of central human functional capabilities, and a broader objection to some elements of the suggested list.

There are evident overlaps between this list and Bailey's serving competencies, as well as notable differences. Among the latter, one needs to be addressed. Bailey selected logical reasoning among the competencies he deemed necessary for a certain kind of liberal education. I have instead maintained reasoning as subsumed in literacy and numeracy, thus presenting it contextualized in terms of discursive and logical reasoning. Furthermore, I have included the capability of practical reason in terms of relating means and ends and evaluating actions, thus implying a form of logical reasoning, albeit more morally oriented. This is a debatable position, since some educationists argue that learning reasoning skills has to be done per se, as well in association with other skills. However, at a basic level the reasoning entailed by literacy and numeracy, as well as by other capabilities such as sociality and practical reason seems to respond adequately to the task of identifying educational enabling conditions.

Comparing this list of basic educational capabilities with Nussbaum's central human functional capabilities (2000, pp. 78–79) appears also interesting and shows consistent overlaps between the two. In particular, two insights derived from considering Nussbaum's "Senses, Imagination and Thought" and "Practical Reason" are significant in this context. The first relates to the similarities between capabilities constitutive of "Senses, Imagination and Thought" and the capabilities selected at an educational basic level. More specifically, Nussbaum refers to capabilities such as imagining, thinking, reasoning, as well as using imagination and thought in relation to self-expression. Moreover, Nussbaum connects these capabilities and their cultivation to the role of an adequate education, consisting of, albeit not limited to, "literacy and basic mathematical and scientific training" (2000, p. 79). The second insight relates to the understanding of practical reason and more generally to the possibility of including capabilities concerned with the acquisition and development of a conception of the good as well as of a critical reflection on one's life. As we have seen, I have included practical reason in terms of reflecting on the relation between means and ends and on actions, rather than in the more substantive terms of forming a conception of the good in relation to one's life. This latter understanding appears as pertaining more to higher levels of capabilities, both in terms of educational and general capabilities. However, the level identified can be considered a first initiation to the more demanding aspect required by forming a conception of the good. This is certainly a debatable issue and my discussion in this context, as I said previously, aims primarily at identifying questions rather than providing a comprehensive account.

There is a consistent objection that can be raised with respect to the subsets of enabling conditions identified as constitutive of the capability to be educated. This relates to the possible understanding of the elements of the list as expression of a "dominant" conception of education. For instance, literacy may be considered "elitist" in the sense of reflecting the norm of the dominant social, cultural, and institutional arrangements. The objection relates also to the possible use of literacy, or any other element of the list, as a means of discrimination and reproduction of structural inequalities,[7] that is, as a means of excluding groups of people who do not share the same language, culture, or class. The answer to this objection rests on providing a justified account of the kinds of opportunities for functionings required to be participant members of one's social arrangements. Literacy, in the specific, plays a substantial role in our society, and its inclusion in the subsets of enabling conditions appears necessary precisely in order to avoid inequalities. The "specific" form that literacy will take is evidently context-dependent, and more complex social and economic arrangements will require, and therefore privilege, certain forms of literacy rather than others. Furthermore, it is perhaps worth reiterating here the importance that literacy has in changing and improving the lives of those who have been marginalized and excluded, as attested both by empirical studies and by more theoretical ones.

Finally, there are important reasons as to why this subset of capabilities for educational functionings constitutes an entitlement and its provision becomes a matter of justice. As we have seen, the capability to be educated relates to the need of education in order to avoid harm or disadvantage to the individual. It is also fundamental and foundational to different and future capabilities. Absence or lack of education would essentially harm the person, or put the person at a considerable disadvantage both for present and for future capabilities. The capability approach identifies the space of capabilities as the relevant one for seeking equality. In other words, equality has to be sought in the substantive freedom people have to choose the life they value, hence in their substantive well-being freedom. Since education plays a crucial role in people's well-being, it follows that unequal opportunities or access to education and its fundamental enabling conditions would constitute an unacceptable inequality. The fundamental capability to be educated and its constitutive elements are crucial parts of the demands of equality, and as such, they constitute a fundamental educational entitlement.

Ultimately, conceptualizing the capability to be educated differs consistently from other views. What is important in capability terms is not simply the amount of resources spent on education, or a consideration of education as a resource in itself, for instance. It is not even the simple production of educational "outputs" in terms of qualifications and years of schooling, as evaluated by economic approaches to education. Rather, the capability approach, as noted in the introduction to this chapter, requires focusing on the contribution that the basic capability to be educated makes to the formation and expansion of human capabilities, and hence to the contribution it makes to the opportunities people have for leading flourishing lives. The fundamental educational functionings and capabilities identified, in promoting human agency, knowledge, and skills, as well as the ability to deliberate over means and ends and basic conditions of autonomy, are all crucial to that important aim.

A concluding comment concerns the provisional nature of my account. This conception of the basic capability to be educated, and the subset of capabilities for educational functionings that ensues, are intended as very preliminary and subject to discussion and revision. I maintain that the fundamental role education plays in people's well-being demands further and more extensive analysis of these and related questions.

Acknowledgment

I am grateful to Harry Brighouse and Terry McLaughlin for their helpful insights, and to the editors of this collection for their encouragement and suggestions. I wish to thank the Philosophy of Education Society of Great Britain for generous and sustained research funding. An earlier version of this chapter was presented at the Fourth International Conference on the Capability Approach, September 5–7, 2004, Pavia, and at the meeting of the Capability and Education Network in Cambridge, November 2004.

Notes

1. Sen does not identify nor defend a specific list of basic capabilities. However, he mentions some basic freedoms, which are seen as fundamentally important to human well-being. This issue is addressed also in a further section of the chapter.
2. The literature on feral children highlights the harm caused by the absence of social interactions and various forms of learning (see, among others, Candland 1995; Lieber 2002). (See also Lane 1977; Newton 2003).
3. Martha Nussbaum (2000) has notably addressed this issue within the approach by articulating a perspective that explicitly formulates a list of central human functional capabilities considered fundamental to human beings. Critiques external to the capability approach point out how the absence of specific capabilities in Sen's approach leads to several theoretical as well as more practice-oriented problems, such as, for instance, that of identifying which capabilities, among the countless ones that people may have reason to value, should receive institutional attention in just schemes.
4. For a clarification of this theoretical problem, see Crocker (1995, p. 154).
5. These further criteria are proposed and discussed by Robeyns in her account of relevant capabilities for gender inequalities (see Robeyns 2003, p. 78).
6. I owe this suggestion to Terry McLaughlin.
7. Perspectives in sociology of education relate the use of hegemonic forms of culture and education to the oppression and subordination of groups of people, for example, disabled people, or people in lower socioeconomic classes (among the vast literature on this aspect, see, for instance, Ball 2006, 2003; Freire 1972; Oliver and Barnes 2002).

References

Alkire, Sabina. 2002. *Valuing freedoms: Sen's capability approach and poverty reduction.* Oxford: Oxford University Press.
Bailey, Charles. 1984. *Beyond the present and the particular: A theory of liberal education.* London: Routledge and Kegan Paul.
Ball, Stephen, J. 2003. *The more things change: Educational research, social class and 'interlocking' inequalities.* London: Institute of Education.
———. 2006. *Education policy and social class.* London: Routledge.
Brighouse, Harry. 2000. *School choice and social justice.* Oxford: Oxford University Press.
Candland, Douglas K. 1995. *Feral children and clever animals: Rflections on human nature.* Oxford: Oxford University Press.
Crocker, David 1995. Functionings and capabilities: The foundations of Sen's and Nussbaum's development ethics. In *Women, culture and development: A study of human capabilities,* edited by M. Nussbaum and J. Clover. Oxford: Clarendon.
Ferri, Elsa, John Bynner, and Michael Wandsworth. 2003. *Changing Britain, changing lives: Three generations at the turn of the century.* London: Institute of Education.

Freire, Paulo. 1972. *Pedagogy of the oppressed.* Harmondsworth: Penguin.

Lane, Harlan. 1977. *The wild boy of Averon.* London: Allen and Unwin.

Lieber, J. 2002. Nature's experiments, society's closures. *Journal for the Theory of Social Behaviour* 27 (2): 325–343.

Newton, Michael. 2003. *Savage girls and wild boys: A history of feral children.* London: Faber and Faber.

Nussbaum, Martha C. 2000. *Women and human development: The capabilities approach.* Cambridge: Cambridge University Press.

————. 2004. Women's education: A global challenge. *Signs: Journal of Women and Culture in Society* 29 (2): 325–355.

Oliver, Michael, and Colin Barnes. 2002. *Disability studies today.* Cambridge: Polity Press.

Robeyns, Ingrid. 2001. *Understanding Sen's capability approach.* Available from http://www.ingridrobeyns.nl (accessed January 7, 2006).

————. 2003. *Sen's capability approach and gender inequality: Selecting relevant capabilities.* Available from http://www.st-edmunds.cam.ac.uk/vhi/nussbaum/papers/robeyns.pdf (accessed January 4, 2005).

Saito, Madoka. 2003. Amartya Sen's capability approach to education: A critical exploration. *Journal of Philosophy of Education* 37 (1): 17–33.

Schuller, Tom, John Preston, Cathie Hammond, Angela Brassett-Grundy, and John Bynner. 2004. *The benefits of learning: The impact of education on health, family life and social capital.* London: Routledge Falmer.

Sen, Amartya. 1980. Equality of what? In *The Tanner Lectures on Human Values,* edited by S. McMurrin. Cambridge: Cambridge University Press.

————. 1992. *Inequality re-examined.* Oxford: Oxford University Press.

————. 1999. *Development as Freedom.* Oxford: Oxford University Press.

Swift, Adam. 2003. *How not to be a hypocrite: School choice for the morally perplexed.* London: Routledge.

Unterhalter, Elaine, and Harry Brighouse. 2003. Distribution of what? How will we know if we have achieved Education for All by 2015? Paper read at the Third International Conference on the Capability Approach: From Sustainable Development to Sustainable Freedom, September, Pavia, Italy.

Williams, Bernard. 1985. The standard of living: Interests and capabilities. In *The Standard of Living,* edited by A. Sen. Cambridge: Cambridge University Press.

3

Situating Education in the Human Capabilities Approach

Pedro Flores-Crespo

Key Issues

It is commonly agreed that education has been intimately connected to capabilities. However, several questions remain unanswered about the way in which knowledge promotes the real freedoms that people enjoy. The aim of this paper is thus to *situate* education in the human capabilities approach. This argument unfolds in three stages. The first will discuss how, historically, education has been regarded as a force capable of enriching the lives of individuals. The second will explain how both Amartya Sen's and Martha Nussbaum's approaches complement each other, allowing a deeper understanding of the links between knowledge and capabilities. Despite this, a more comprehensive approach is needed. Therefore, in the third stage, in order to examine to what extent education contributes to the expansion of capabilities, it is necessary to look at four dimensions: (1) philosophical (how education modifies human agency or "personal autonomy"); (2) pedagogical (how knowledge is constructed/transmitted); (3) institutional (how schools in particular and education system in general function in order to enable the expansion of freedoms); and (4) policy issues (how the educational problem is formulated and how policy outcomes can be evaluated).

Introduction

Sen has made a remarkable contribution to social sciences and humanities by pioneering the capability approach. He has led us to reexamine contemporary social theory by arguing that we need to look at what people are able to do or to be, when making normative evaluations. This contrasts with measuring people's feelings or material possessions. Without denying the importance of income or commodities, Sen has helped us to "direct our attention" to other important types of information when we talk about development, well-being, and quality of life (Des Gasper 1997).

The notion of capability basically consists in identifying people's functionings, which represent the achievements of a person: that is, what the person manages to do or to be (Sen 1985). These achievements may vary, as Sen argues, from elementary (being adequately nourished and being free from avoidable disease) to "very complex" activities or "states of existence" (being able to take part in the life of the community and having self-respect). According to Sen:

> A person's capability refers to the alternative combinations of functionings that are feasible for her to achieve. Capability is thus a kind of freedom: the substantive freedom to achieve alternative functioning combinations (or less formally put the freedom to achieve various lifestyles).
>
> (1999, p. 75)[1]

Although Sen's capability approach has had a considerable impact on empirical social-science literature, as Alex Callinicos (2000) notes, in the field of education there are still several questions that remain unanswered. For instance, is education still related to freedom as classical thinkers remarked? In which way has the enlargement of capabilities been analyzed from an educational perspective? Is it sufficient to claim that by receiving formal or informal education, a person expands automatically her or his freedoms? Which other factors could intervene in the expansion of capabilities of an academically trained individual? Addressing these four questions is the primary aim of this chapter. What I therefore hope to do is to open up a reasoned discussion on how education can be situated within the capability approach.

Education and Humanity A Brief Historical Account

> The agents of destiny are men, and men conquer freedom when they are aware of their destiny.[2]
>
> (Octavio Paz)

Historically, education has been intimately connected to human capabilities. From Socratic times until our day, an array of voices has lucidly explained how knowledge helps us to clear our minds, awaken our consciousness, inform our actions, and enrich our lives. According to Nussbaum, the Stoics, for example, argued that the central task of education was to confront the passivity of the pupil, challenging the mind to take charge of its own thought. The ancient Greek philosophers claimed, she argues, that "people who have conducted a critical examination of their beliefs about what matters will be better citizens—better in emotion as well as in thought" (Nussbaum 1997, p. 29). During the Renaissance, Desiderius Erasmus, a man profoundly influenced by the classical philosophers, returned to these classical ideas to defend the view that reason ought to be the driving force of human nature and that the function of education was to enable human beings to enjoy life to the full. As a humanist, Erasmus called for an end to cruel practices commonly employed in sixteenth-century schools and rejected rote learning. There is no reason, he said, to make the school experience a tough and joyless process (Curtis and Boultwood 1966). Two centuries after Erasmus, the intrinsic relationship between education and humanism was profoundly enriched by the work of some of the greatest thinkers. John Locke, for instance, the precursor of liberalism, published in 1693 *Some Thoughts Concerning Education* (Locke 1693), where he explained that although nature might be a persistent factor in the formation of human beings, being academically trained could help a pupil to learn how to use reason in making decisions. Curtis and Boultwood have argued that Locke's general principle lies in the assumption that "true education is the gaining of self-control by the individual" (1966, p. 243). In other words, education is a process of the formation of personal autonomy.

Locke's ideas found opposition in the writings of philosopher Jean-Jacques Rousseau, who published *Émile* in 1776. In his novel he described how human beings should be taught within a corrupt world. Dave Robinson and Oscar Zarate argue that *Émile* was a "revolutionary" book since it examined the whole concept of childhood, stressing the necessity to stimulate compassionate feelings by means of education. Émile's "nature feelings of pity," needed to be transformed into "imaginative empathy" for others in order to become in a "truly virtuous" citizen of the world (Robinson and Zarate 2001, p. 70). For Rousseau, being educated implied becoming a "loving and tender-hearted"[3] individual, capable of understanding why others succumb unintentionally to miseries. It was necessary "to perfect reason through feeling," he said (Robinson and Zarate 2001).

Influenced by Rousseau, Immanuel Kant, the German philosopher, reflected on whether people are by nature, morally good or bad. He concluded that "he [sic] is neither, for he is not by nature a moral being. He only becomes a moral being when his reason has developed ideas of duty and law" (Curtis and Boultwood 1966, p. 296). Kant thought that as human beings are rational creatures, they could learn through nurture, discipline, moral training, and instruction. This is not to say, however, that human beings are free from having inclinations to vices. Kant, as opposed to Rousseau, recognized that people may have natural inclinations to every vice and that those inclinations and instincts could urge them one way. Notwithstanding, the German philosopher also noted that reason could, however, drive them in another direction. Education, therefore, should aim at promoting reasoning abilities in order to distinguish between virtues and vices and then be able to *act* accordingly.

If reason can be achieved by means of education, and if this faculty serves to better human beings in an ethical way and to enrich our lives, then it can be said that education and human freedom have been inextricably linked across the ages. Following these philosophers, reason would then seem to be a central element to situate education within the capability approach. Both Sen and Nussbaum have stressed the importance of reason in the expansion of capabilities. On the one hand, Sen (1992) speaks of "reason to value," which means that people need to scrutinize their motivations to value specific lifestyles. On the other hand, Nussbaum, considers "practical reason" as one of the central capabilities for functioning and suggests that practical reason and affiliation[4] have special importance since they both "organize and suffuse" all other capabilities (Nussbaum 2000, p. 82). Seeing reason given a predominant role in the process of capability enlargement—and as a capability in itself—leads us to ask under what circumstances education could privilege reasoning abilities. I will return to this point again later.

Apart from redirecting our attention to consider reason as a driving force of capabilities, the historical account given above leads us to the claim that the notion of capabilities developed by Sen and Nussbaum recaptures the humanistic view of education, which had been partially eclipsed by simplistic application of modernist theories of education (see, for example, human capital theory, manpower planning, or the idea of a knowledge-based economy[5]). The restoration of the humanistic vision to the academic debate constitutes a good reason to engage in the analyses concerning the links between education and freedom (understood as capability). In doing so, previous approaches and various authors need to be reexamined in order to understand how education contributes to the expansion of human freedom. In this sense, consider, for instance, the great pedagogue Paulo Freire's *Education: The Practice of Freedom* (1972) and *The Pedagogy of the Oppressed* (1967). In the latter book, Freire pointed out that

> Education as the practice of freedom—as opposed to education as the practice of domination—denies that man [*sic*] is abstract, isolated, independent, and unattached to the world; it also denies that the world exists as a reality apart from people. Authentic reflection considers neither abstract man nor the world without people, but people in their relations with the world.

> (1972, p. 62)

Freire, Nussbaum, and Sen conceive human beings as responsible agents who can alter their destiny. While Freire argued that education becomes the means by which people can perceive, interpret, criticize and, eventually, transform their reality, Sen and Nussbaum have stressed the importance of individual agency and practical reason in the process of enlarging people's freedoms.[6] So, by studying Freire's, Sen's, and Nussbaum's perspectives, our understanding of how a pedagogical practice can bring wider benefits can be expanded. As we can see, the links between education and freedom are not novel; rather, what has changed is the way of analyzing this connection.

Possibilities, Uses, and Gaps of the Capability Approach in Education

(i) What might be the merits and the weaknesses of the capability approach applied within the educational field?
In her analysis, Elaine Unterhalter (2001) argues that education appears undertheorized in Sen's human capability approach. She rightly notes that in the formulation of his approach, Sen considers education as a: (i) social opportunity, (ii) valuable outcome (literacy skills), and (iii) causality of freedom. But she questions the mere fact of widening educational opportunities because this process does not always go in the same direction as the process of expanding human capabilities. Unterhalter explains that, despite increasing student enrolment ("social opportunities," in Sen's terms[7]), the human capabilities of women are far from being expanded in many education systems. To illustrate this, she provides as an example the educational system of South Africa, in which a high proportion of young black girls by attending school, are at risk of being sexually harassed and assaulted by their teachers and even by their schoolmates. Under such an "unregulated social facility," she concludes, the so-called academic environment can literally end a girl's life through HIV infection, thus destroying her capability (2001, p. 7). Unterhalter continues with her critique of the capability approach by saying that Sen's approach entails "two different senses of education that are sometimes confused" (2001, p. 5). First, education can be seen as a form of functioning or well-being achievement, for example, completing four years of basic education in school. Second, education can also be thought of as part of the process of exercising agency, that is, using reflection, information, understanding, and the recognition of one's right to exercise these capacities in order to formulate the "valued beings and doings" (2001, p. 5). This is what I would call reason, which seems to be the underexplored element in analyses of capability achievement and education.

Nonetheless, Unterhalter's claims are very useful because they help us to avoid simplistic applications of the capability approach and urge us to think more broadly about the educational process and the supporting conditions that should exist within an academic context. We cannot assume that education promotes valued doings and beings automatically. But, equally, it would be foolish to overlook the other side of Unterhalter's argument, that the three aspects of education shown in the capability approach (social opportunities, schooling, and the development of reason) can complement each other in order to achieve valuable capabilities. If a person has an equal educational opportunity, the person's practical skills and human agency can be shaped in a fair way.

(ii) An application in education research
In my own applied study (Flores-Crespo 2002) I sought to provide a comprehensive explanation of the relationship between education and development in Mexico by using an evaluative framework. This framework was formed by two key elements: first, a list of seven functionings for university graduates that were defined by following Nussbaum's list of central capabilities, and second, two "instrumental freedoms" proposed by Sen (see table 3.1). Instrumental freedoms (or "capability enhancers") contribute, as Sen (1999) argues, directly or indirectly to the overall freedom that people have to live the way they would like to live. He identifies five distinct types— political freedoms, economic facilities, social opportunities, transparency guarantees, and protective security. Surprisingly, some educational commentators have not properly discussed the role of these means of development in their analyses (see, for example, Unterhalter 2001; Saito 2003).

In methodological terms, it is worth mentioning that three technological universities were selected as case studies, and their graduate functionings were evaluated by means of a questionnaire.[8] The survey was followed by qualitative data collection. More than a hundred semistructured interviews were conducted with employed and unemployed graduates, employers, and educational officers with the aim of extending the structured data gathered through the questionnaire.

My research showed that most graduates achieved valuable personal and professional achievements thanks to the education provided by the selected universities.[9] That is, education had a positive impact on these graduates' functioning capabilities. Nevertheless, through a comparative analysis between the three cases, it was also found that graduates' capabilities varied significantly as a consequence of preexisting circumstances. In short, as instrumental freedoms vary, so does individuals' functionality. I argued that the impacts of education can be potentially maximized if

Table 3.1   Proposed framework to evaluate the capabilities of university graduates

Proposed functionings	Sen's instrumental freedoms	Nussbaum's central capabilities
Personal achievements ("beings")		
1. Being able to feel confidence and self-reliance		Being able to avoid unnecessary and nonbeneficial pain, so far as possible, and to have pleasurable experiences
2. Being able to visualize life plans	Social opportunities and economic facilities	Being able to form a conception of the good and to engage in critical reflection about the planning of one's life ("practical reason")
3. Being able to develop further abilities	Social opportunities and economic facilities	Being able to think and to reason and to do these things in a way informed and culti- vated by an adequate education
4. Being able to transform commodities into valuable functionings	Economic facilities	
Professional achievements ("doings")		
5. Being able to acquire knowledge required in a job position	Social opportunities	Being able to think and to reason and to do these things in a way informed and cultivated by an adequate education
6. Being able to look for and ask for better job opportunities	Economic facilities	Being able to move from place to place Being able to form a conception of the good and to engage in critical reflection about the planning of one's life ("practical reason")
7. Being able to choose desired jobs	Economic facilities	Being able to form a conception of the good and to engage in critical reflection about the planning of one's life

Source: Adapted from Flores-Crespo, 2002.

social opportunities and economic facilities are generated simultaneously and, conversely, a lack of these instrumental freedoms will constrain educational endeavor.

Although the merits of this study should be subjected to a wider discussion, it can be said that my analysis highlighted the multidimensionality of the capability approach.[10] That is, human well-being is multidimensional, and it should advance many different kinds of capabilities at the same time. It was shown that professional achievements and even monetary benefits overlapped with personal functionings. For example, a female graduate argued that going to university helped her find employment, which was crucial for earning a wage that could be used for improving her life as an independent person. Otherwise, she said, "I would be a housewife only devoted to looking after many babies" (Flores-Crespo 2002, p. 262).

Another feature of the capability approach that I demonstrated was the necessity of broadening the informational basis when speaking about the "virtuous cycle" between education and economic growth or development. Being academically trained, employed, and relatively well paid does not necessarily imply *development*. It was found that, although most graduates received an income, they were still facing other type of inequalities (in Sen's words, "unfreedoms") such as long and exhausting shifts (in some cases, illegal) and gender discrimination during the process of personnel selection and hiring. Moreover, the lack of broader educational opportunities in the most disadvantaged localities (Tula and Nezahualcóyotl) halted the aspiration of many graduates to continue their professional studies beyond the technical degree. As technicians in Mexico are relatively badly paid and do not have high social status, the majority of graduates wished to study further for a bachelor's degree. However, such reasoned aspiration could not be met since the supply of tertiary education in poor areas is still insufficient. Families cannot afford the cost for their young people to travel to other cities to study, and there were no institutional arrangements to upgrade technical expertise into bachelor degrees within the higher education system of Mexico.

Lastly, I will comment on two general limitations of this study although more voices should also undertake this critique. First, my study says little about social stratification and, therefore, further work needs to be done in this regard. Although university graduates gain a valuable personal position in comparison to their counterparts who did not study, it is not clear to what extent someone born in a relatively deprived area and who attended a technical college could achieve a middle- or upper-class position. In order to clarify this point, a broader sociological approach is needed. This confirms Robeyns's (2005) claim that applying the capability approach to issues of social change will often require the addition of explanatory theories. Second, a more precise way of defining functionings is needed in order to choose relevant variables that, eventually, could be included in the survey. The question is how the list of functionings shown table 3.1 is a complement to the established learning objectives of the selected universities. It is important to stress here the fact that functionings and capabilities are broader (and more complex) achievements than a learning outcome. So, the process of expanding people's choices through education must be initiated by accomplishing the academic objectives included in a course. This will demand a fruitful combination of pedagogical and institutional factors within schools in particular and education systems in general. I will return to this point later (see (ii) and (iii) in the next section).

In summary, my work sought to mark a difference from common analyses that normally relate education to economic development by considering educated people as an end of the development process and not just a means or instrument of the progress, as suggested instead by the human capital perspective. In doing so, Sen's definition of human development and his capability approach were very illuminating.

(iii) Present and future freedoms

Madoka Saito (2003) is another commentator who has contributed to the discussion of the capability approach in the field of education. In her analysis, she raises two interesting points, among others, which help us to reflect more deeply on the future possibilities of Sen's capability approach. First, she asks, "how can we apply the capability approach to children, since children are not mature enough to make decisions by themselves?" (2003, p. 25). Sen addressed this question by saying that "when you are considering a child, you have to consider not only the child's freedoms now, but also the child's freedom in the future" (in Saito 2003, p. 25).[11] Furthermore, according to Sen, his approach can be applied to children since we can judge their well-being in terms of their functionings (what they can actually do) rather than their mental attitudes. But, in addition, "the freedom aspect may be important for a child because: firstly, a child makes some decisions, like whether he or she is being unhappy, wants milk and so on; and, secondly a child's future involves the time when the child will actually exercise some freedom" (Sen in Saito 2003, p. 26). Saito concludes that "as long as we consider a person's capabilities in terms of their life span, the capability approach seems to be applicable to children" (2003, p. 26).

Second, Saito points out that "the kind of education that best articulates the concept of Sen's capability approach seems to be the one that makes people autonomous and, at the same time, develops people's judgement about capabilities and their exercise" (2003, p. 29). I agree with Saito, though

I would add that the type of education that seeks to encourage the achievement of personal autonomy usually is known as "liberal education."[12] So, if education needs to be coherently situated in the capability approach, it is necessary to comment on the foundations of such a type of education.

Liberal Education and the Capability Approach: Toward a Renewed Perspective

Liberal education is so called because it seeks to liberate the mind from the bondage of habit and custom, according to Nussbaum (1997). This type of education, she argues, is based on the assumption that human beings should not be directed by other voices nor become instruments on which fashion and habit play their tunes, since they possess an innate capacity for reasoning and questioning. According to Alan Ryan, an education can be identified as liberal, if it is aimed at preparing autonomous, argumentative, and "tough-minded individuals" who eventually will become "good liberal citizens" (Ryan 1999, p. 84). Reason, then, appears to be a precondition of personal autonomy and, thus, both elements form part of the concept of liberal education.

But, once having shown that reason, personal autonomy, and independence[13] are elements that help us situate education in the capability approach, we confront other problems: How can reason and, thus, personal autonomy be achieved? Are there some pedagogical guidelines to make liberal education a reality in times of "economic anxiety?"[14] It is fair to recognize that the capability approach can enable the revival of liberal education in that it stresses reason and personal autonomy?[15] Despite this possibility, it seems that there are several elements that need to be discussed in order to more broadly understand the role of education within the capability-based approach.

(i) The philosophical side: The search for substantive freedoms
Speaking of liberal education awakes an old debate: Is this type of education totally incompatible with vocation-oriented training? Are the aims of liberal and vocational education completely irreconcilable? Christopher Winch and John Gingell (1999) point out that there are different ways of categorizing the aims of education. One tradition, they explain,

> emphasizes the importance of education as an individual, liberal good with intrinsic values. Another tradition sees education as a public, as well as an individual good, with instrumental, as well as, or in contrast to, intrinsic value. Broadly speaking, the former tradition is called *liberal,* the latter *instrumental.* Instrumental aims can be further classified into vocational, societal and personal.
>
> (1999, p. 11, authors' emphasis)

The insightful distinction between the two purposes of education has been obscured by the belief that these aims are mutually incompatible or in constant rivalry. Ryan (1999), for example, says that the sharpest challenge to the liberal ideal of education is posed by a commercial and vocational ideal of student competence. On the other hand, Winch (2002) notes that some philosophers have failed in recognizing the significant aspect of vocational education. It is assumed, as Winch and Gingell note, that there is a "complete dichotomy" between vocational and liberal education. However, they argue, this is a fallacy because although two things may not share all the same characteristics, they certainly can have something in common. Vocational educators "may espouse liberal aims like personal fulfillment, while liberal educators may espouse vocational aims like employability as a form of personal fulfillment" (Winch and Gingell 1999, p. 247). This view was supported in part by the results of my empirical study (Flores-Crespo 2002), where an overlap between personal and professional achievements was clearly identified.

As the capability approach is increasingly studied and applied in the field of education and, consequently, the revival of liberal education seems to be at least a possibility,[16] the reconciliation of intrinsic and instrumental aims of education is needed to make wise use of Sen's and Nussbaum's ideas. I would also claim that promoting personal autonomy through liberal education does not necessarily imply a sense of detachment from the reality (or that it is exclusively for the middle and upper classes). On the contrary, liberal education endorses the view that human beings must think and act in a complex environment, and thus, they must be equipped with relevant knowledge and independent-minded attitudes to deal with such complexity. In this sense, Freire wrote in *Education: The Practice of Freedom* that only by developing a permanently critical attitude could people overcome a posture of adjustment in order to become integrated to reality (Freire 1973). Here, it is important to remember that Freire's ideas have been used and applied in contexts where people are facing several disadvantages that need to be overcome. So, speaking on freedom or promoting it among poor people is neither an extravagance nor an unrealistic attempt and education can make the difference when its intrinsic and instrumental values are wisely combined and taught.

As I have argued, Freire's ideas bear a close resemblance to those of Sen and of Nussbaum. The pedagogue, the economist, and the philosopher both support and depart from the assumption that as people are succumbing to poverty, inequality, exploitation, and ignominy, something needs to be done to reverse this situation. These authors do not seem to be trying to "save" the poor as they conceive human beings as responsible agents who can alter their destiny. So, Freire, Sen, and Nussbaum rely upon the human agency of individuals to transform their realities. This leads us to a stimulating discussion concerning the way in which formal or informal education could shape our human agency in order also to exercise it ethically. It is precisely in the connection between substantive freedom and education that educational commentators need to focus further attention if we want to advance in the understanding of the capability approach and its relation to education.

(ii) The pedagogical side: Under what conditions liberal education could be realizable

> A bad education is one in which compassion does not fit, which, carried out by the cult of rationality, pretends that human existence be completely intelligible and ignores contradictions.[17]
>
> *(Pablo Latapí Sarre)*

Nussbaum, unlike Sen, has provided quite certain normative pedagogical guidelines on how to promote freedoms by means of education (specifically, higher education). According to Nussbaum (1997), three *capacities* are essential to the "cultivation of humanity" in today's world. First, is the capacity for critical self-examination. The Stoics, according to the American philosopher, claimed that people who conduct a critical examination of their beliefs about what matters will be better citizens—"better in emotion as well as in thought" (1997, p. 29). Argument helps students not only to clarify their ideas, but also to act rationally in one way rather than another. The second capacity refers to the necessity of conceiving ourselves as citizens of the world. That is, to adopt, as educators and as students, an attitude of mutual respect, "we must educate people who can operate as world citizens with sensitivity and understanding," says Nussbaum (1997, p. 52). This capacity needs to be nourished in higher education in the classroom itself as well as in its reading material. Here, the role of educators is also crucial because we need to show our students,

> the beauty and interest of a life that is open to the whole world, to show them that there is after all more joy in the kind of citizenship that questions than in the kind that simply applauds, more fascination in the study of human beings in all their real variety and complexity than in the zealous pursuit of superficial stereotypes, more genuine love and friendship in the life of questioning and self-government than in submission to authority.
>
> (Nussbaum 1997, p. 84)

This capacity has particular relevance in contexts where large minorities and indigenous groups try to coexist with the majority. Intercultural practices are perfectly entailed in this capacity. It is important to remember that the goal of education, according to the Stoic norm of world citizenship, is not the separation of one group from another, but respect, tolerance, and friendship (Nussbaum 1997, p. 67). But, despite the value of this capacity, reality is more complicated. Phenomena such as bullying, racism, homophobia, or gender discrimination can easily penetrate school, college, and university walls, affecting the humanity of specific groups of individuals. I will return to this point later.

Lastly, the third capacity proposed by Nussbaum, which is closely related to the first two, is narrative imagination. "This means the ability to think what it might be like to be in the shoes of a person different from oneself, to be an intelligent reader of that person's story, and to understand the emotions and wishes and desires that someone so placed might have" (Nussbaum 1997, pp. 10–11). This capacity can be encouraged by means of arts, specifically literature. "Higher education should develop students' awareness of literature in many different ways," since it will foster an informed and compassionate vision of the different (Nussbaum 1997, p. 88). Compassion, she adds, requires a sense of one's own vulnerability to misfortune.

Nussbaum's normative ideas on higher education, as well as her list of central capabilities, represent a step forward in the design of "just pedagogies which can be tested and adjusted empirically," as Melanie Walker (2002) notes. Unlike utilitarian "key skills conceptions of team work," Nussbaum's capabilities offer a comprehensive idea for a pedagogy of inclusion, according to Walker (2002, p. 5). Highlighting the possibility of the capability approach in pedagogical terms requires us to look at other aspects, such as how knowledge is provided. In this sense, educational contents, written and visual materials, commonly used in the classroom, can present images or assertions that are far from Nussbaum's idea of cultivating humanity. For example, by examining illustrations, images, and language of didactical instructional material for science courses, Mary T. Hardy (1989) found a manifestation of gender bias in the science curriculum in primary texts, particularly those associated with the physical sciences. She noticed that males were more frequently represented in cards of such material than females. Women's images were often derogatory (doing silly things), and boys' interest was appealed to more than girls'. This could explain, according to Hardy, the shortage of girls taking up the science option. This pedagogical error shows, on the one hand, that although educational opportunity may exist, gender inequity can prevail. On the other hand, it shows that, in the future, girls may have less freedom to choose an academic course in science. Ingrid Robeyns confirms this point by saying that in England "career aspirations are still highly gendered, and boys are frequently found to be dominating the classroom environment and monopolizing teacher's time" (2002).

Apart from the necessity of designing appropriate material to encourage pupils to learn in an environment of mutual respect, it is clear that further work needs to be done to understand the process of capabilities expansion from a pedagogical perspective (but see Walker 2005). With the aim of contributing to this debate, I will raise the following questions:

- Do learning aims say something useful for the design of valuable functionings? If so, could a curriculum be based on functionings rather than on educational objectives? Will there be an innovative approach in doing so?
- "Should we generate a core curriculum in education in order to enhance the capabilities in children?" (Saito 2003, p. 29).
- Do skills, competence and professional functionings mean the same? Can learning outcomes be deemed as a capability?
- Is it illogical to think that the results of academic assessment processes could help us evaluate functionings (consider, for example, the Programme for International Students Assessment of the OECD)?
- How can we put into practice "narrative imagination?"

As we see, several issues are waiting to be investigated to grasp a coherent picture of how education promotes capabilities.

(iii) The institutional side: Schools as capabilities-oriented institutions?

In order to define a framework that may help us to situate education in the capability approach, it is necessary to look at schools from an institutional side. Schools in particular, and education systems in general, constitute social settings where changing patterns of behaviors, customs, and values take form and are reproduced. Some of these habits, irrespective of the educational supply, could affect both pupils' development of reason and their acquisition of skills. In this regard, Lucy Trapnell (2003) reports that in Peru, as schooling was expanded, a reverse effect was generated. While the Ashaninka people thought that school education would help improve their life conditions, they became aware that school education was not offering them the opportunity they had hoped for. This was because

> their children were taught everything from a western world view, but were actually acquiring very little western knowledge. When they finished primary school the majority could hardly read or write, and lacked basic mathematical skills. Many had a very limited command of Spanish—yet their own indigenous culture was being ignored, even vilified. As a result, children grew up with negative attitudes towards their cultural heritage and with a very low sense of self-esteem.

(2003, p. 9)

But expressions of superiority are not exclusive to developing countries. Human Rights Watch (HRW)[18] has documented the "devastating impact of pervasive animus" toward lesbian, gay, bisexual, and transgender youth (between the ages of 12 and 21) in seven states of the United States. Each day, most gay youth walk into their schools wondering what they will have to face—taunts, food thrown in the face, lewd mockery in the locker room, being slammed against lockers or walls. "Not surprisingly, they lose their focus; their grades drop, some drop out, and a few commit suicide" (Human Rights Watch 2001, p. 174). For HRW, the public school system in the United States is failing systematically to protect these students. It adds:

> The burden these students bear is exacerbated in many cases by the rejection of their families, condemnation within their communities, being demonized by individual teachers and administrators, and rejection by members of the adult lesbian, gay, bisexual, and transgender communities who are too scared of being identified themselves to offer support to gay youth.
>
> (2001, p. 4)

Academic environments are not immune to family prejudices, social indifference, and impunity for school officials, and this urges us to pay careful attention to how educational institutions function within their particular contexts. More specifically, we need to pay careful attention to how homophobia; bullying;[19] discrimination, whether based on gender, identity, sexual orientation, race, ethnicity, immigration status, economic situation, or disability; and violence have been engendered and reproduced in educational systems.

Looking at the institutional side of the learning process might be useful since our understanding of the way in which educational processes can expand (or restrict) our freedoms is broadened. We should not overlook that, as Sen reminds us, individuals live and operate in a world of institutions and that our "opportunities and prospects depend crucially on what institutions exist and how they function" (1999, p. 142).

(iv) Policy cycles

Public policies are a response to public problems. So, when a government acknowledges the existence of a public problem and the need to do something about it, policymakers need to decide on some course of action (Howlett and Ramesh 2003). According to Unterhalter, governments that adopt the capability approach also have "an obligation to establish and sustain the conditions for each and every individual, irrespective of gender, ethnicity, race or regional location, to achieve valued outcomes" (Unterhalter 2003, p. 4). This claim raises the normative position that can underpin a capability-based policy. However, in addition to policy formulation, and even more important, is the need for a theory of justice that should "tell us whether a policy is liable to promote, be consistent with or violate justice, what ought to be" (Brighouse 2002, p. 1). Although the capability approach can serve as an important constituent for a theory of justice, as Robeyns remarks, it does not amount to a theory of justice. As Robeyns notes, Sen himself has made clear that the capability approach specifies an evaluative space (individual freedom), which, as was said earlier, has been a remarkable contribution to social sciences.

The capability approach, according to Robeyns (2005), is a "broad normative framework for the evaluation and assessment of individual well-being and social arrangements, the design of policies, and proposals about social change in society." It provides, she argues, a tool to conceptualize and evaluate social phenomena such as poverty, inequality, or well-being (2005, p. 94). Thanks to this evaluative capacity, the freedom-based perspective can be useful for reformulating public actions and for correcting policy failures that may lead to inequality, exclusion, or poverty. In the education field, this can be achieved more easily once careful attention is paid to the pedagogical practices and the institutional side of schools and educational systems. Revising teaching models and institutional settings could allow us to take measures in order to counteract the effects of oppressive cultural manifestations and regressive social mores.[20] This task can be undertaken and reinforced by considering "instrumental freedoms" as a central element of Sen's approach. Instrumental freedoms, as Sen argues, contribute directly or indirectly "to the overall freedom people have to live the way they would like to live" (1999, p. 38).

Furthermore, focusing on both functioning achievement and instrumental freedoms leads us to have a more comprehensive and *realistic* vision of what education can do for human beings since we are evaluating not only the substantive freedoms gained by academic instruction but also the social, economic, and cultural circumstances that condition the educational endeavor. This approach, as it was argued elsewhere, contrasts to the functionalist perspective that considers education as the "engine" of development or economic growth (Flores-Crespo 2005, 2002). Education needs friendly conditions in order to expand human capabilities, and public policies can work substantially in creating such conditions.

Concluding Remarks

In order to situate education within the capability approach, it is necessary to recognize the simple but important fact that the educational process normally occurs within institutions, that knowledge is achieved through written and visual material, that pupils are guided by teachers, and that generally students are educated by having an intense social interaction with others. The school environment, therefore, entails diverse factors that may condition the acquisition of knowledge, the development of reason and, therefore, our present and future human freedoms. Education can certainly contribute to the expansion of capabilities, but, under certain conditions, it can also function with the opposite result.

In analyzing the expansion of capabilities from an educational angle, I have therefore suggested a four-dimensional framework, which is shaped by these elements: a philosophical, a pedagogical, an institutional, and a policy side. Through this heuristic exercise it is possible to think more carefully about a renewed concept of liberal education and about previous approaches that have lucidly associated education with freedom. Arguably, there are important lessons to be drawn from this literature.

Furthermore, the proposed pedagogical normative ideas can be considered as a step forward to inform didactic practices in the classroom. Notwithstanding, several questions remain open on curriculum design, on the appropriate use of educational jargon, and how people learn. Clearly, much work needs to be done in this sense.

Looking at the institutional side of the educational systems helps explain why the capabilities of an academically trained person can be expanded or restricted. This may encourage a reflective move by society and governments and, consequently, capability-based policies can be formulated. As we can see, there are many issues waiting to be investigated to make education a *real* driving force for the enlargement of human freedoms.

Acknowledgments

Thanks to Melanie Walker for her useful comments on earlier drafts. I am grateful too to the members of the Education and Capability Network and to participants of the seminars organised at Cambridge, Bath and York in May 2005 for insightful discussions. Thanks too to Deborah Fabri and Catherine Fanning for their reading and to Dulce Mendoza for helping me with the editing.

Notes

1. On the foundation bases of the capability approach see Sen 1985, 1999 and Nussbaum 2000 (see also Alkire and Black 1997; Cameron and Gasper 2000 for a critical review). Stimulating points of view can be found in the archive *The Capability Approach* at www.fas.harvard.edu/~freedoms/ca_conference_fourth.html.

2. "*Los agentes del destino son los hombres y los hombres conquistan la libertad cuando tienen conciencia de su destino.*" (Translated by Séverine Deneulin.) This is a fragment of the opening speech delivered by the Nobel laureate at the conference, El siglo 20: La experiencia de la libertad. Mexico, October 8, 1990. *Vuelta* magazine, 167.

3. These terms belong to Jacques Rousseau (in Curtis and Boultwood 1966, p. 282).

4. This refers to (a) being able to live with and toward others, to recognize and show concern for other human beings, to engage in various forms of social interaction, to be able to imagine the situation of another, and (b) having the social bases of self-respect and nonhumiliation; being able to be treated as a dignified being whose worth is equal to that of others (Nussbaum 2000).

5. A critical review of these approaches can be found in Flores-Crespo 2004, 2002.

6. It is worth commenting that some authors, rather than focusing on convergences, have contrasted Freire's approach to Sen's. Raff Carmen (2000, p. 1023), for instance, criticizes Sen's capability approach due to its "individualistic mindset" and "abstract formalisms." He points out that the freedom-based perspective cannot be operationalized and capability is described in "static (non-process) terms."

7. Social opportunities, according to Sen, "refer to the arrangements that society makes for education, health care and so on, which influence the individual's substantive freedom to live better" (1999, p. 39).

8. The higher education institutions selected were (1) The Technological University of Nezahualcóyotl in the state of Mexico, (2) The Technological University of Tula-Tepeji in the state of Hidalgo, and (3) The Technological University of Aguascalientes in the state of Aguascalientes. The sample was made from 717 graduate questionnaires that were distributed proportionately in the three localities.

9. The technological universities were created with the primary purpose of offering educational services to those people who had been excluded from higher education. Because of this, some of these colleges are situated in relatively poor areas (e.g., Tula-Tepeji and Nezahualcóyotl).

10. I thank Dr. Flavio Comim for drawing attention to this point.

11. According to Sen, "The main argument for compulsory education is that it will give the child when grown up much more freedom and, therefore, the educational argument is a very future oriented argument" (Sen in Saito 2003, p. 27).

12. We should remember that personal autonomy is the fundamental value underlying all other claims under different versions of liberalism, according to Harry Brighouse (2002).

13. In this sense, Dearden (1972) raises an interesting question: "Is autonomy the same thing as freedom, or perhaps independence?" He explains that all three are closely interrelated, but the three are distinguishable. Freedom, he argues, consists in the absence of constraints or restraints relevant to what we do want or might want to do, and it is not a sufficient condition for saying that a person is autonomous. "A person might be given more freedom in his job, but lack the relevant knowledge, skills, initiative, experience which are necessary for an appropriate exercise of autonomy in this enlarged role [. . .]". Autonomy,

meanwhile, "is invulnerable to the constraints or restraints of others" and an autonomous agent must be independent-minded, though some kinds of independence will be matters of our external relations to people and things (Dearden, 1972, pp. 450–452).

14. This expression belongs to Martha Nussbaum (1997). She observes that many institutions that call themselves liberal art colleges have adopted a vocational orientation.

15. We also need to remember that both personal autonomy and liberalism perspectives are grounded in individualism, understood as "an outlook which assigns primacy to individual human beings," according to the *Penguin Dictionary of Philosophy* (2000), not as an egocentric position.

16. Interestingly, the International Commission on Education for the Twenty-first Century, led by Jacques Delors, has emphasized that "One of education's principal functions is [. . .] that of fitting humanity to take control of its own development. It must enable all people without exception to take their destiny into their own hands so that they can contribute to the progress of the society in which they live [. . .]" (Delors 1996, p. 78–79). Even some international financial agencies such as the World Bank have shifted, as Lynn Ilon (1996) noted, their rationales concerning educational policies: now they focus more on welfare issues rather than on productive purposes.

17. *"Mala es una educación en la que no cabe la compasión; mala la que, llevada por el culto a la racionalidad, pretende que la existencia humana sea cabalmente inteligible e ignora sus contradicciones"* (Latapí 2001). Translated by Séverine Deneulin. Pablo Latapí Sarre is a professor at the National Autonomous University of Mexico (UNAM), and currently, ambassador of Mexico to UNESCO.

18. Human Rights Watch was founded in 1978 as an independent, nongovernmental organization, supported by contributions from private individuals and foundations worldwide. It defends freedoms of thought and expression, due processes, and equal protection of the law, and opposes arbitrary imprisonment, discrimination, and other abuses of internationally recognized human rights. Its goal is to hold governments accountable if they transgress the rights of their people.

19. According to Ken Rigby, bullying is a repeated oppression, psychological or physical, on a less powerful person by a more powerful person or group of persons (Rigby 1996, p. 15). It is widely prevalent in schools, and it can be greatly reduced, if not entirely eliminated, principally by actions taken by schools and also, to a lesser degree by parents. The most common types of bullying are physical (hitting, beating, kicking, spitting, throwing stones) and psychological (verbal abuse, name calling, threatening gestures, stalking, malicious telephone calls to a student's house, repeatedly hiding another's belongings, leaving people out of desired activities, and spreading malicious rumors about someone).

20. Aimed at reducing bullying at schools, the project Kids Help Line (KHL) was implemented in Australia as a free service for children with problems who want to talk to a counselor by telephone. It is worth saying that during a twelve-month period, KHL received over 7,000 calls from children about bullying (Rigby 1996).

References

Alkire, Sabina, and Rufus Black. 1997. A practical reasoning theory of development ethics: Furthering the capabilities approach. *Journal of International Development* 9 (2): 263–279.

Brighouse, Harry. 2002. Egalitarian liberalism and justice in education. Inaugural Professional Lecture, Institute of Education, University of London.

Callinicos, Alex. 2000. *Equality.* Great Britain: Polity.

Cameron, John, and Des Gasper. 2000. Amartya Sen on inequality, human well-being, and development as freedom. *Journal of International Development* 12 (7): 985–1045.

Carmen, Raff. 2000. Prima mangiare, poi filosofare. *Journal of International Development* 12 (7): 1019–1030.

Curtis, Stanley James, and Myrtle Emma Amelia Boultwood. 1966. *A short history of educational ideas.* (4th edition reprinted). Great Britain: University Tutorial Press.

Dearden, Robert Frederick. 1972. Autonomy and education. In *Education and the development of reason,* edited by R. F. Dearden, P. H. Hirst, and R. S. Peters. London, Boston: Routledge and Kegan Paul.

Delors, Jacques et al. 1996. *Learning: The treasure within. Report to UNESCO of the International Commission on Education for the Twenty-first Century.* Paris: UNESCO.

Flores-Crespo, Pedro. 2002. An analysis of the relationship between higher education and development by applying Sen's human capabilities approach. PhD thesis, Politics Department, University of York, York.

———. 2004. La relación entre educación y desarrollo bajo una perspectiva crítica. In *Comparative Education and International Education,* edited by L. Lázaro and M. J. Martínez. Spain: Universitat de Valencia.

———. 2005. *Educación superior y desarrollo humano. El caso de tres universidades tecnológicas* Mexico: Asociación Nacional de Universidades e Instituciones de Educación Superior.

Freire, Paulo. 1967. *Education: The practice of freedom.* Translated by Myra Bergman. Great Britain: Hazell Watson and Viney. 1973.

———. 1972. *Pedagogy of the oppressed.* Harmondsworth: Penguin.

Hardy, Mary T. 1989. *Girls, science and gender bias in instructional materials. Series of occasional papers edited by Janet Atkin and Martin Coles.* School of Education, University of Nottingham.

Howlett, Michael, and M. Ramesh. 2003. *Studying public policy. Policy cycles and policy subsystems.* 2nd ed. Canada: Oxford University Press.

Human Rights Watch. 2001. *Hatred in the hallways. Violence and discrimination against lesbian, gay, bisexual, and transgender students in U.S. schools.* USA: Human Rights Watch.

Ilon, Lynn. 1996. The changing role of the World Bank: Education policy as global welfare. *Policy and Politics* 24 (4): 413–424.

Latapí, P. 2001. *Los triunfadores (1289).* Available from www.proceso.com.mx (accessed July 15, 2001).

Locke, John. 1693. *Some thoughts concerning education.* Available from www.socsci. kun.nl/ped/whp/histeduc/locke.

Nussbaum, Martha C. 1997. *Cultivating humanity. A classical defence of reform in liberal education.* Cambridge, MA: Harvard University Press.

———. 2000. *Women and human development:The capabilities approach.* Cambridge: Cambridge University Press.

Rigby, Ken. 1996. *Bullying in schools and what to do about it.* London, Bristol, Pennsylvania: Jessica Kingsley.

Robeyns, Ingrid. 2002. Sen's capability approach and gender inequality. Paper read at Promoting Women's Capabilities: Examining Nussbaum's Capabilities Approach, September 9–10, at Von Hügel Institute, St. Edmund's College, University of Cambridge, UK.

———. 2005. The capability approach: A theoretical survey. *Journal of Human Development* 6 (1): 93–114.

Robinson, Dave, and Oscar Zarate. 2001. *Introducing Rousseau.* Australia: Icon Books.

Rousseau, Jean-Jacques. 1776. *Émile.* Available from http://www.ilt.columbia.edu/ pedagogies/rousseau/index.html.

Ryan, Alan. 1999. *Liberal anxieties and liberal education. What education is really for and why it matters.* Great Britain: Profile.

Saito, Madoka. 2003. Amartya Sen's capability approach to education: A critical exploration. *Journal of Philosophy of Education* 37 (1): 17–33.

Sen, Amartya. 1985. *Commodities and capabilities.* Amsterdam: New Holland.

———. 1992. *Inequality re-examined.* Oxford: Oxford University Press.

———. 1999. *Development as freedom.* Oxford: Oxford University Press.

Trapnell, Lucy. 2003. Identity crisis. *Developments: The International Development Magazine, Glasgow:* DFID 22 (2nd quarter): 9–11.

Unterhalter, Elaine. 2001. The capabilities approach and gendered education: An examination of South African complexities. Paper read at the First International Conference on the Capability Approach: Justice and Poverty, Examining Sen's Capability Approach, September, St Edmund's College, University of Cambridge, UK.

———. 2003. Education, capabilities and social justice. In *Chapter prepared for UNESCO EFA Monitoring Report.* London.

Walker, Melanie. 2002. Gender justice, knowledge and research: A perspective from education on Nussbaum's capabilities approach. Paper read at Promoting Women's Capabilities: Examining Nussbaum's Capabilities Approach, September 9–10, at Von Hügel Institute, St. Edmund's College, University of Cambridge, UK.

———. 2005. *Higher education pedagogies: A capabilities approach* Maidenhead: SRHE/ Open University Press and McGraw-Hill.

Winch, Christophe, and John Gingell. 1999. *Key concepts in the philosophy of education.* London: Routledge.

Winch, Christopher. 2002. The economic aims of education. *Journal of Philosophy of Education* 36 (1): 101–117.

4

Distribution of What for Social Justice in Education? The Case of Education for All by 2015

Elaine Unterhalter and Harry Brighouse

The chapter considers how assessments are made concerning global social justice and education. It explores the need to consider not only international patterns of access to and very narrowly defined achievements in education but also to assess the distribution of other aspects of education deemed valuable, particularly given complex global class, gender, race and ethnic inequalities. It uses the example of the movement to achieve Education for All (EFA) by 2015 associated with UN agencies, international nongovernmental organizations (NGOs), and the work of many governments worldwide, and looks critically at the indicators that have been developed to evaluate this. We argue these are inadequate to capture what would rightly count as "education for all." We argue instead for the appropriateness of the capability approach, particularly taking issue with its Rawlsian critics such as Thomas Pogge, who defend a resourcist alternative. We highlight how thinking about distribution and capabilities requires a range of different ways of evaluating social justice in the provision of education. This opens up institutions to critique and we offer some thoughts toward how this approach to evaluation might be developed to assist in critical policy work.

In 1990, at the Jomtien Conference in Thailand, organized by UNESCO, UNICEF, UNDP, and the World Bank, 157 governments agreed to the World Declaration on Education for All that signaled their commitment to achieve EFA by 2000. EFA was not defined succinctly, but was laid out as comprising universal access to education services "of quality"; equity with regard to removing disparities "in access to learning opportunities" for certain groups, such as girls, women, "the underserved" and the disabled; and learning acquisition and outcome in "useful knowledge, reasoning ability, skills and values." The acquisition of learning in a range of different settings was acknowledged, but emphasis was placed on primary schooling for children of appropriate age (World Declaration 1990). By 2000, despite some large-scale initiatives by some governments, and the emergence of NGOs as important participants in the movement for EFA, the goals set in 1990 had not yet been realized. A further meeting at the World Education Forum in Dakar in 2000 led to governments and NGOs from 164 countries agreeing to a Programme of Action to implement the Jomtien Declaration. At the Millennium summit of the UN, two Millennium Development Goals (MDGs) were set in education: gender equity in education by 2005 and EFA by 2015. This chapter considers not so much whether the targets associated with these goals will be met, nor whether targeting is an appropriate approach in education. An extensive literature on these issues already exists (e.g., Black and White 2003; UNESCO 2003a). We are concerned with a different order of question, namely, how we will know whether we have social justice in education or EFA by 2015. The chapter draws on insights from the capability approach to critique some of the indicators and other measures being used to assess EFA. In the concluding section it offers some thoughts toward the development of an alternative form of evaluation grounded in the capability approach, considered both on its own terms, and in relation to a number of criticisms of the approach.

The chapter is divided into three sections. The first outlines the existing measures used to evaluate progress toward the MDGs and offers some critique of their assumptions, scope, and reliability. We highlight how these measures do not address the full range of concerns represented by the concept "education for all" and point out how the capability approach critiques assumptions entailed by these measures. The second section assesses some difficulties with the capability approach, but suggests that the approach is at least as good as its main rival. In the final section of the chapter we build from our defense of the arguments in the capability approach to offer some suggestions regarding how a different evaluative measure for EFA might be developed.

Approaches to Measuring Education for All

The three most widely used measures relating to EFA are the gross enrolment rate (GER), the net enrolment rate (NER), and the gender gap. The GER is the proportion of all children of school-going age attending school on a given census day. The GER is often over 100 percent because in schools in many countries there are large numbers of underage children (because of inadequate preschool provision) and overage children (because of high levels of repetition). The GER is sometimes disaggregated by gender and district. But the GER cannot measure whether the children in school on the census day attended regularly or only for that day. The NER is a measure of whether children of the appropriate age group (6–11) are in primary schools. The NER is considered a much more refined measure of EFA than GER, and is again often expressed in disaggregated form by district or gender. However, the NER, like the GER, cannot measure whether children attend regularly or not, or their responses to learning and teaching in the school. In addition, in many countries children's births are not registered. In these circumstances there is no accurate data on their age on entry into school and hence the NER is likely to be based on guesswork. The GER and NER are sometimes supplemented by a measure of the gender gap, that is, the ratio of enrolment of girls to boys. The gender gap can be a measure of enrolment or completion. But as a measure of gender equality it is very limited as it only gives information about the levels of girls' enrolment or attainment relative to boys' and does not consider wider meanings of gender inequality within school or as a result of school (Unterhalter, Rajagopalan, and Challender 2005). In recent years the Global Monitoring Reports published by UNESCO have also compiled data on the intake ratio of children to school, repetition rates, survival rates, numbers of teachers, levels of adult literacy, and expenditure on schooling (UNESCO 2005, pp. 251–421).

The collection of data to calculate the NER, gender gaps, school progression, and repetition rates has been subject to scrutiny over the past decade since it became evident that the Education Management Information Systems (EMIS) in many countries were rudimentary. Often decisions of considerable moment were being taken with inadequate data about the numbers of teachers or children in school and a lack of time series data on which to base predictions about supply and demand (Heynemann 1997; Samoff 1994; Carr-Hill et al. 1999).

Throughout the 1990s, the dissatisfaction of governments, development assistance organizations, and NGOs with the imprecision of enrolment-based measures of EFA led to a push for more output-oriented measures, primarily test scores, and more complex assessments of school resources. For the Dakar meeting on EFA UNESCO commissioned EFA assessments, which were tests in language, arithmetic, and life skills carried out on a representative sample of children in a large number of countries in year 4 of school. The tests were intended to assess learning achievements, and generally found very low levels in the children tested in developing countries against national levels of achievement and in cross-national tests (UNESCO 2000). The extent to which test scores provide an evaluation of learning, rather than just what children have been taught to demonstrate for the test is a hotly debated area (Goldstein 2004; Wolf 2001). Tests cannot avoid the problem that some children might not demonstrate in the test what they know, some learning may not be amenable to testing in this way, and decontextualised measures of achievement, without any understanding of school contexts, are ever open to misinterpretation. In addition, test scores are subject to the problem of aggregation (even when they are disaggregated by gender or location). They provide information about how a school, a district, or a country performs, but very rarely is this information used in relation to any other information concerning the social background of an individual child.

The limitations of test scores have resulted in attempts to model more-complex measures of educational quality and equality in developing countries. Kevin Watkins developed the Education Performance Index (EPI) (Watkins 2000). The EPI comprised measures of coverage based on the NER, completion, data on the proportion of learners who progress beyond year 4, and gender equity, based in turn on the gender gap in primary schooling. The EPI ranks each of these measures against a perfect score for that measure (e.g., 100 percent progression beyond grade 4). This is set alongside a measure of average income levels in a country using World Bank data on purchasing power parity that expresses the number of units of a national currency required to purchase a basket of goods that could be bought for $1 in the United States (Watkins 2000, pp.131–141). Watkins acknowledged the difficulties relating to the quality of data for calculating the NER and accepted there were problems with regard to seeing completion rates as a means to indicate learning, and viewing gender inequality only in terms of the gender gap in enrolment. His work was one of the first attempts to integrate different forms of measurement of education quality and consider how these linked to national income. However it is noticeable that for all the achievements of this measure, it provides only country-level rankings and little information regarding what aspects of education quality are valued in particular settings, and how aggregated country data can provide insights into complex forms of inequality. Later modifications of the EPI in the UNESCO Global Monitoring Reports through the development of the Education for All Development Index (EDI) and the Gender-related EFA Index (GEI) continued to focus on measures of inputs (enrolment rates), outputs (survival over five years in school), and parity (equal numbers of girls and boys at different levels of schooling), but did not take further the social justice question of what the distribution of education was for (UNESCO 2005, p. 418). The limited form of measurement entailed a rather limited form of policy, that is, the expansion of schooling generally paying insufficient heed to questions of quality and equality in schools. The numbers of teachers, textbooks, or schools were considered appropriate proxies for measuring these (UNESCO 2003b).

Efforts to understand education quality in ways that go beyond enrolment, and that give insights with regard to local level settings have taken three directions. First, a number of researchers have tried to develop measures of education quality based on data collected at the household level or through participatory processes in communities. The demographic and health surveys (DHS) that have been conducted in 47 developing countries since 1992, with an attempt at representative sampling, provide information on individuals and the numbers of years in school. The surveys are based on data collected from households and not school data. The latter is sometimes of doubtful value because local education officials may believe that false data showing higher enrolment might better secure their job. Data collected at the household level might also be fabricated, in that household members might report more or fewer years in school than have in fact been accomplished. But they generally have no compelling incentives to do this.

Second, participatory studies based in communities have been used to consider the views of the very poor and assess some of the reasons they consider education valuable (Alkire 2002; Young 2006; Nelson Mandela Foundation 2005). One of the issues all these studies, which pay considerable attention to local context and meanings, encounter is how these insights can be generalized across different populations to help with guiding global education policy.

A third approach to assembling local level data and developing forms of measurement linked to this has focused on the school. This form of measurement has affinities with ideas of school effectiveness and new public management. Some measurement builds from detailed qualitative observations of learners and teachers, for example, the model outlined by Helen Craig and Ward Heneveld (Craig and Heneveld 1996). This has been utilized in a number of different settings beyond the African countries in which it was initially developed. For example, in Bangladesh, the model has been used in the Primary School Performance Monitoring Project linked to the government's concern with improving quality, and detailed research has examined some of the effects of the form of teacher interactions with learners (Unterhalter, Ross, and Alam 2003; Ferdous and Rahman 2000). But some trenchant critiques have been made of school-based indicators of quality with regard to what information they provide on equality. A study of indicators used in the Basic Primary Education Programme in Nepal highlighted confusion and overlap in the indicators collected (Singh et al. 2002). A number of studies in the UK show how monitoring school-level information on children's achievement in tests or the proportion of children receiving free school meals fails to provide a full picture of the complex forms of inequalities and engagements to change this (Arnot and Reay 2006; Lupton et al. 2006). In South Africa, the allocation of school budgets on the basis of distance from a tarred road, access to electricity, indoor or site taps, an office for the principal, entails that schools, which have these amenities, but which enroll children from very low-income families, will have similar budgets to schools where children come from high-earning families (Unterhalter 2006). While school-based data, like that gathered from households, overcomes some of the problems of data quality and aggregation linked to country-level data on enrolments, survival, and parity, it is no better at suggesting what indicators point to social justice in education and not just resources for education.

Some education economists have addressed this issue by modeling schools in relation to inputs and outputs and developing a measure of education production function. This approach develops measures of student attainments or school outputs that are seen to be a function of student inputs (say, baseline test scores, parental level of education and involvement with children's education, peer group effects) and aspects of school context that can be quite elaborate and score school climate, leadership style, learner motivation (Levacic and Vignoles 2002). In some contexts, for example, the UK, studies of this nature are possible because detailed data is available on individual children, and instruments to measure school context have been developed. A number of internal critiques are made of this model: for example, the problem of two-way causality built into it and the fact that it only takes account of endogenous measures. However, over and above these critiques, is the assumption that an input-output model is an appropriate model of schooling and of education; that socioeconomic data on children (for example, parents' income or education level) provides enough information to take on board assessment of complex problems of inequality (particularly the more qualitative dimensions), and that the complex processes of school context can be scored through crudely developed instruments.

An attempt to bring together approaches that looked in detail at school resources, learner performance, and human rights emerged in the South African government's *Report to the Minister: The Financing, Resourcing and Costs of Education in Public Schools* (South Africa 2003). This report was concerned with developing instruments to ensure interprovincial and intraprovincial equity in resourcing, noting that education personnel were more equitably distributed within provinces than learning outcomes (in tests). The report recommended investigating a subsidy for nonpersonnel costs linked to poor learners across the country, whether or not they are in "poor" schools. It also supported the development of an integrated performance monitoring system open to public scrutiny and showed the relationship between levels of resources and the socioeconomic status of a district and learner performance. This was an important attempt to link socioeconomic information, concern with human rights, pro–poor policies, and public accountability that goes well beyond any of the existing approaches to measurement. However, problems concerning reliance on test scores as a measure of learner performance remain. This report raises, but does not solve, issues with regard to how accountability to local demands for education in terms of human rights are negotiated and how the impetus in the approach toward rights-based measures might articulate with other concerns relating to efficiency and fiscal discipline. Shireen Motala, in commenting on the report, pointed up the need, given historic inequities, for greater specificity regarding how a level of adequacy of resources and minimum conditions for learning are defined (Motala 2003, pp. 9–10). Thus, while the review is sensitive to a wide range of issues often ignored in discussions of funding formulae, the process of redistribution was bedeviled by only preliminary means to conceptualize redistribution and equality.

So current approaches to measuring EFA are problematic. They assume that education means either enrolment in school, or achievement in a narrowly constructed test. Gendered identities are treated as descriptive categories of biological difference between children. They disregard concerns of gender analysis of inequality and debates concerning the intersectionality of gender and other forms of social division (Robeyns 2006; Anthias 1998; Phillips 1999). The goal of EFA is universal inclusion in *quality* primary education. There is considerable debate about how quality in schooling is to be defined, but this debate rarely takes account of the how quality in education links to equality in a society (UNESCO 2005, p. 36). While the quality debate is sometimes settled according to various understandings of school effectiveness or school inputs, very little attention is given to how the question of education equality is to be settled.

We think that the capability approach has enormous potential to address these problems of equality, quality, and measurement. The capability approach alerts us to the need to describe not only access to, and very narrowly defined achievement in, education but also to assess aspects of education deemed valuable and hence issues about the distribution of resources, given complex class, gender, race, and ethnic inequalities.

The Capability Approach

The capability approach, as developed in the work of Amartya Sen, provides a very useful way, given the complexity of diverse societies in the world, to think about social justice, and especially gender equality, in education. Sen (1999) takes issue with the approaches to evaluating social policy that focus on the aggregated benefits an initiative has for the whole society or for future generations, without regard to the distribution of effects on individuals. According to these views, for example, investing in education for women and girls is justified by its benefits not for them, but for the societies they live in. These approaches do not look at whether any adult or child has been discriminated against in the provision of education, because the education is not for those individuals but for a larger grouping—the community, then nation, and future generations. These views might be weakly interested in gender equality in education, but only insofar as it is needed to ensure a range of social benefits.

The capability approach looks at a relationship between the resources people have and what they can do with them. As Sen puts it, in a good theory of well-being, "account would have to be taken not only of the primary goods the persons respectively hold, but also of the relevant personal characteristics that govern the conversion of primary goods into the person's ability to promote her ends"(Sen 1999, p. 74). What matters to people is that they are able to achieve actual functionings, that is, "the actual living that people manage to achieve" (Sen 1999, p. 73). The concept of functionings reflects the various things a person may value doing or being varying from the basic (for example, being adequately nourished) to the very complex (for example, being able to take part in community life). But interpersonal comparisons of well-being should incorporate references to functionings, and also reflect the intuition that what matters is not merely achieving the functioning but being free to achieve it. So we should look at "the freedom to achieve actual livings that one can have a reason to value" (Sen 1999, p. 73). A person's capability refers to the alternative combinations of functionings that are feasible for the person to achieve. Capability is thus a kind of freedom: the substantive freedom to achieve alternative functioning combinations (Sen 1999, p. 75). The notion of capability is essential for Sen, because someone's actual functionings need not tell us very much about how well off the person is.

Defending Capabilities

Thomas Pogge (2003) has recently criticized the capabilities approach. He presses two major objections: first, the capabilities approach faces a serious problem in dealing appropriately with natural inequalities; second, he thinks it tends to obscure the degree of unjust international inequality. We shall sketch, and briefly respond to, his arguments.[1]

One of the apparent advantages of the capability approach over its rivals is its sensitivity to inequalities of natural endowments. The value of resources is usually defined without regard to what their holder can do with them; but the capability approach always looks at how well an individual can convert her bundle of resources into functionings. Resourcist approaches regard an ordinarily abled person as equally well off as a paraplegic as long as they have the same level of resources. But the capability approach counts the paraplegic as worse off (from the point of view of justice). Pogge claims this is actually a *drawback* of the capability approach. Pogge thinks that many natural inequalities are, in fact, horizontal—of a kind with differences in hair color, skin color, height, and other features:

> While the resourcist approach is supported by this conception of natural inequality as horizontal, the capability approach requires that natural inequality be conceived as vertical. When a capability theorist affirms that institutional schemes ought to be biased in favour of certain persons on account of their natural endowments, she thereby advocates that these endowments should be characterized as deficient and inferior, and those persons as naturally disfavored and worse endowed . . . not just in this or that respect, but overall.
>
> (2003, pp. 54–55)

How can the capability approach respond to this objection? One response would be to criticize resourcism for insensitivity to the fact of impairment. But resourcism can acknowledge that some of the functionings unavailable to disabled people are unavailable not because they suffer from physical impairments, but because social institutions are set up so as to enhance the functioning of the ordinarily abled but not the disabled person. Blindness does not make the written communication of others inaccessible to someone, the absence of Braille facilities does.

But if Pogge is right that many disabilities result from social institutional bias rather than directly from impairment, the capability approach can agree. It can attribute the failure to function adequately to the institutions, and say that this implies that they should be reformed. It does not make a fetish of correcting the individual rather than the institutions.

But what if there really *is* inequality of capabilities—if those who are "disabled" really cannot reach the same level of functionings as others even if there is reform of institutions? Then Pogge's objection loses much of its power. Resourcism cannot straightforwardly compare the blind person and the sighted person and say that they have unequal shares of resources; they do not. In order to locate the disadvantage socially, and avoid stigmatizing the blind, it has to say instead that the blind person is disadvantaged by the design of social institutions. How is it going to establish that disadvantage? The blind person does not have an expensive taste, as, for example, we might think of someone who is sighted but prefers reading Braille to reading print, because she enjoys the tactile experience. But why not? It's hard to explain why not without appealing to the fact that she (unlike the sighted Braille reader) lacks a valuable capability absent but for the provision of Braille. It is hard to see how the primary goods approach can determine whether social institutions are set up to the disadvantage of the disabled without appealing to some notion of functioning.

Second, think more centrally about the way that girls and boys can face similar resources in schools; these similar resources can give rise to differential opportunities because of their different needs. A school without running water is inconvenient for everyone, but much more so for girls, who menstruate, than boys, who do not. Even absent social mores that look down on menstruation, many girls may find that without adequate water provision they cannot attend school on the days that they menstruate heavily. There is nothing wrong with the girls; acknowledging that their different physicality gives rise to different needs does not imply any stigma. Girls in many schools in South Africa face a high risk of rape at school because the security arrangements are so poor (Unterhalter 2003). The security arrangements are poor for all, but if a girl is raped and becomes pregnant, this severely diminishes her access to future educational opportunities. Again, acknowledging this difference does not imply stigmatization.

Does the capabilities approach downplay the level of global inequality? Pogge writes:

> Consider how the scores for the US and India are calculated. One begins with their raw per capita GDPs of $34,737 and $453. One then adjusts both amounts by valuing the two relevant currencies at purchasing power parity (PPP) rather than market exchange rates, yielding $34,142 for the US and $2,358 PPP for India. The final step converts these numbers into logarithms, yielding (after normalization) 0.97 for the US and 0.53 for India. Through these two steps, an initial inequality ratio of 77 is reduced to 14.5 and finally to 1.83.
>
> Compare this new HDI metric with the old metric of national per capita incomes. Here is a holdover from the past: "The income gap between the fifth of the world's people living in the richest countries and the fifth in the poorest was 74 to 1 in 1997, up from 60 to 1 in 1990 and 30 to 1 in 1960. [Earlier] the income gap between the top and bottom countries increased from 3 to 1 in 1820 to 7 to 1 in 1870 to 11 to 1 in 1913." Crude as it is, this resourcist statement provides crucial information. It says a lot about the avoidability of poverty: One percent of the incomes of the people in the rich countries would suffice to increase the incomes of those in the poorest countries by 74 percent. It says a lot about the distributions of bargaining power and expertise, which condition international negotiations and agreements. And it says a lot about

how successive global institutional schemes distribute the benefits of global economic growth.

(Pogge 2003, p. 217)

Notice that Pogge's objection is really to the Human Development Index (HDI), rather than to Sen's more theoretical rendering of his view; it could be the index, rather than Sen's view, that is at fault. Assume, though, that the HDI is a more or less faithful representation of the capability approach. We agree that the poor of the world are much less well off than the rich of the world, and any measure that failed to acknowledge this would merit no further consideration. But as a given society's holdings of wealth increase, the opportunities for well-being (however that is understood, except in terms of wealth, obviously) do not increase in a linear fashion, or even in a fashion that is well understood.

Consider the longitudinal evidence within distinct economies. Between 1972 and 1991, real GDP per capita grew (pretty steadily) in the United States by 39 percent. The percentage of respondents to polls reporting themselves as "very happy" barely increased at all during the same period; and the kinks in that curve bear no relationship to the steady rise in the growth curve (Frank 1999, p. 72). In Japan, GNP per capita grew steadily from 1960 to 1987 by a total of 300 percent; the average reported level of well-being reported by respondents to surveys changed barely at all year to year, hovering around 6 (out of 10) (Frank 1999, p. 73). These findings chime with Fred Hirsch's argument in *Social Limits to Growth* that, past a certain point of material development, as the material economy grows, the positional economy becomes an increasingly dominant part of the material economy (Hirsch 1977). The positional economy relates to certain kinds of goods that cannot be more widely distributed because their value lies in the social construction of their high status, and part of that status rests on the fact that access to these goods is limited. Places at elite higher education institutions are examples of positional goods, as is work in high-status occupations or access to very select forms of leisure. Growth in the positional economy shows up as a growth in wealth, but it does not bring any contribution to overall well-being, and so should not be counted when we are comparing how well off people are from one society to another.

The problem that positionality poses for resourcist metrics is that it makes it unclear how to compare the real resource base of people in separate societies that differ with respect to the extent to which positional goods dominate their economies (Brighouse and Swift 2006). The worries posed by consideration of positional goods and other similar goods that relate to status or identity, in that their relationship to income is not linear may seem to be of limited practical importance in the context of developing countries. After all, whatever the consequences of considering the complexities introduced by positional goods, the poor in the developing world are much poorer, and much worse of, than most people in the developed world. But they are important for the task of reconsidering the criteria for EFA.

Different countries, even at similar levels of development, will distribute education (which is in part a positional good) differently. Even for different countries with the same distribution of education it will be more or less positional (and positional in different ways) depending on both the distribution of labor market rewards and the extent to which education influences labor market rewards. So concerns about positionality affect even comparisons among developing countries, especially in educational contexts.

Toward an Alternative Approach to Assessing EFA

In conceptualizing measurements of EFA that take account of a wide conception of education and some of the debates about equality, we now suggest some combinations of the primary goods and capability approaches. The diagram (figure 4.1) below expresses our thinking.

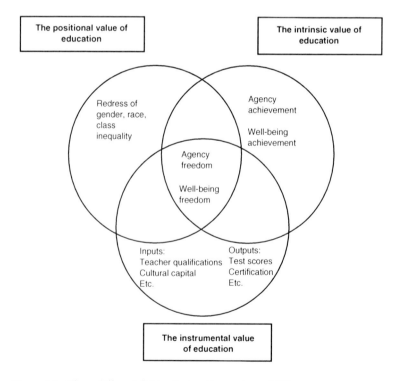

Figure 4.1   Three different fields of education and capabilities

Before discussing the model in depth we want to offer some responses to criticisms of the attempt to develop any approach to measurement. Some critics point out that any development of indicators is likely to suffer from the criticisms we have outlined in the first section , that is, poor quality data, crude instruments, inappropriate levels of aggregation, and lack of public accountability. We accept that there is considerable problem in overcoming the difficulties with data quality. However, the crudeness of the existing instruments does not mean we should not attempt to try to improve these and in ways that are more publicly accountable than heretofore. At present it is generally the case that the form of the indicator for EFA, with its stress on enrolment or achievement in narrowly defined tests, drives the policy of governments, NGOs, and intergovernment organizations (IGOs). A more multidimensional indicator might make for more nuanced policy.

We are not suggesting that this or any other model for refining the instruments for measuring EFA should supplant the insights generated by qualitative research. But we do wish to utilize some of the complexities that qualitative researchers alert us to in developing a model that is more sensitive to context than those used heretofore, but also simple enough to be operationalized in a range of different settings.

The model we are developing is very preliminary, but is an attempt to depict some of the thinking linked to the capability approach in relation to education. We have conceptualized education in relation to three different fields. We have identified these as the instrumental value of education, the intrinsic value of education, and the positional value of education. Our conception applies to many different facets of education. For example, passing a course at a certain level has instrumental value in securing access to a particular job; it has intrinsic value in affirming a valued outcome, and it has positional value possibly in identifying and recognizing achievement on a positional scale with regard to class, gender, or race. Similarly, a curriculum might have aspects that stress the instrumental aims, the intrinsic aims, and the positional aims of education.

The three fields that intersect with this terrain of freedom relate to three different understandings of education. First, education has an instrumental value. In this field education (very often understood as schooling) helps secure work at a certain level and political and social participation in certain forms. Without some formal level of skill acquisition, schooling for a number of years or other form of initiation into a group (for example, through learning a sacred language or particular religious practices), one cannot achieve vital aspects of agency and well-being, that is, live a life one has reason to value. Of course, what one needs to learn in order to support the interest in having a life one has reason to value will be somewhat dependent on context. Children in societies with different political economies might need quite different sets of skills. Conditions in formal settings for learning can support the development of the instrumental value of education. For example, the experience of the teacher; adequate time and other resources available for instruction; support for practice of new skills from family, peers, and teacher are all important. While these and the connection of education level with the labor market are quite often measured, the education production function tends to measure only some components of this field, and generally ignores aspects of well-being and agency freedom.

The second overlapping field highlights what we call the intrinsic value of education. This refers to the benefits a person gets from education that are not merely instrumental for some other benefit the person may be able to use it to get. The educated person might have a more rewarding and complex mental life than she had before being educated, regardless of whether the education helps her gain or keep employment. The person might discover the enjoyments attached to reading literature, or appreciating finding out about different forms of music; the person might discover an aptitude and enthusiasm for constructing model figures, which she enjoys even though she is, after the fact, disinclined to sell them. EMIS, DHS, and the education production function do not provide any indicators for thinking about the intrinsic value of education. But it is possible that we could develop proxy measures with regard to what would indicate a gain in agency and well-being achievement, that is, valued aspects of life and success in pursuing valued goals. As in the previous field, consideration of agency and well-being freedoms and the conditions that secure this is part of consideration of the intrinsic value of education. It is here that the model encounters difficulties with regard to the problem of adaptive preference and subjective well-being that has troubled many scholars concerned with refining the capability approach (Gasper 2004). Adaptive preference may mean that a woman values only a very low level of education for herself because that is what is sanctioned by her society. She believes she needs less education than her husband or brother to realize her well-being. However, this field also intersects with the area concerned with agency and well-being freedoms. These are not measured only by subjective preference statements. The field also overlaps with the area of the instrumental value of education, where the extent to which education allows access to the labor market or political participation is measured. Thus the information in this field is gathered multidimensionally in an effort to overcome some of the limitations of studies that have only collected data on reported preference satisfactions.

The third field concerns the positional value of education. Education is positional insofar as its benefits for the educated person depend on how successful she has been relative to others. For example, for any individual child aiming to enter a prestigious university, for which there are a fixed number of places, what matters to her is not at all how successful she has been in school, but only how successful she has been relative to her competitors. If there are 100 places, and she is 99th in line in a mediocre educational regime then she is positionally better off (though worse educated) than if she is 101st in line in a far superior educational regime. Certification (the grades one has achieved) is the most obviously positional aspect of education, but there are many less-flagrantly positional aspects of: schooling. For example, the reputation and location of one's school and whom one attends school with. (It is positionally better to get a mediocre education alongside fellow children of the elite than to get a better education alongside children who are less well connected, even in relatively democratic and meritocratic societies.) Other aspects of the positional dimensions of education are how well the teachers in a school have transmitted "cultural capital" or how far new meanings of gender or race have been developed. Because of the positional value of some aspects of education (very often aspects of a hidden curriculum), schools might have very unequal effects, even though they might look very similar in terms of some measure of education production function. The third field, then, is an attempt to make visible the invisible forms of discrimination by gender, race, or class; forms of misrecognition that occur, and to consider some way of using the school system to effect redress and understanding of past injustice. The level of inequalities in schools could be mapped by using the methodology of the Gini coefficient and scoring other dimensions of inequality (e.g., parental income, education, or other aspects of social identity and their complex intersections). Possibly, proxy measures for a school engaging with forms of redress could be developed. These measures too would need to take account of well-being and agency freedoms and could not be simply handed down from some higher level of the education administration.

At the heart of the three overlapping fields is the concern in the capability approach with well-being and agency freedoms (Sen 1985). These freedoms relate to the social conditions to secure instrumental, intrinsic, and positional values through education. The field of well-being freedom in education is concerned, for example, with freedom from harassment in a classroom, freedom to concentrate in a classroom (not too tired, too hungry, too anxious), freedom to access a lesson through appropriate pedagogies, and good quality of management (Unterhalter 2005). These freedoms apply to adults and children. Agency freedom is concerned with the freedom to access information about education, engage in discussion, and make up one's mind about access to education for an adult without violence or shame. With regard to children, the freedoms entailed concern

freedom from interference with the children's welfare rights and the protection of their potential to develop agency freedom through attending school
(Brighouse 2002; Saito 2003). Placing freedoms at the center of the model is
both an attempt to work with some of the intersections of the capability and
primary goods metrics and an attempt to acknowledge the importance
social conditions have with regard to any instrument of measurement.

The significance of the ideas diagrammed is that measures trying to
gauge whether EFA has been achieved need to reflect these different aspects
of education. We are aware of the need to take account of social conditions
in looking for measures in each field. In some societies with highly central-
ized control of the education system, to achieve aspects of national identity
it could be that the three fields are all suffused by this project and that aspects
of historical context may shape the diagramming in particular ways. There
would be a similar effect in weak or failed states. The flatness of the diagram
does not mean that at certain moments certain fields may not be more
important than others for supporting the interest in living a life one has reason to value. For example, we conjecture that the more unequal a society in
terms of income and wealth, the more important it will be for policy attending to the conditions to realize the instrumental and redress value of education, without neglecting the ways that these both are engaged with
well-being and agency freedoms. However in a society that is more equal in
terms of income and wealth, the instrumental and redress value of education
might concomitantly be somewhat less important than realizing well-being
and agency achievement. And the intrinsic value of education might come
to the fore in those societies; since the structure of the economy and tax-
transfer system contributes more to justice, the education system is freed up
to contribute more directly to well-being.

Well-being and agency freedom stand in the center of all the three overlapping fields. In developing some kind of metric we need to take account
of the fact that much of the work in education, and in meeting the EFA
goals relates to working with children. Thinking through the model (and its
strengths and weaknesses) will entail thinking with processes of change
over a cycle that takes account of children growing up and growing toward
agency freedoms. This aspect of the diagram, responding not only to
changes in historical time, but also to changes over the life of an individual,
raises an added level of complexity. Indeed, it might be that if education is
instrumental in expanding freedoms over a lifetime, the different fields
reconfigure themselves.

Taking on board our critiques of existing measures of EFA outlined in
the first section and our discussion of the capability approach and primary
goods thus far, what are some of the areas we think need to be measured
in order to effect redistribution and an expanded understanding of EFA?

Below is a preliminary, idealized, list. Obviously not all of these factors are readily measurable—the list is generated to demonstrate what we would have to know to be confident about the implementation of EFA. But the other factors *are* readily measurable, and the achievement of EFA should be monitored by appeal to as many of these measures as possible.

i) Conditions for free public discussion of the content and form of education for diverse social groups taking account of complex histories and uneven contemporary patterns of public participation

ii) Conditions for forms of governance regarding education at national, regional, and local level that have been freely agreed to by all adults

iii) Conditions to implement, given a diversity of social settings, the recommendations of public discussions (adequate financial and skill resources to manage and evaluate the implementation, adequate time, legislation for education provision of a particular form enacted under fair conditions)

iv) Processes through which individuals can articulate the intrinsic value of education, drawing on a wide range of forms of expression—freedom of speech on a wide range of topics, access to resources for self-expression

v) Measurements of inequality, such as the Gini coefficients, not just at the level of income and wealth, but also of aspects of identity—race, ethnicity, gender, and their intersections—that might be relevant to forms of inequality. These measures are at the national and local level, including schools

vi) The median income of teachers and the ratio of that figure to the median income in the country, plus noting the extent to which teachers' salaries vary by gender and between teachers of children in higher and lower socioeconomic status

vii) Measures of school input *and* output drawing on a primary goods metric

viii) Measures of resources necessary for adequate achievement of acceptable output linked to participation in the labor market, the political system, and the forms of free public discussion in (i)

Conclusion

We have tried to bring together areas of discussion usually held separate: critical commentaries on indicators of EFA and gender equity in education, an engagement with the justification of the capability approach, and a consideration of how thinking about education as a capability can contribute to

an expanded understanding of education for all. Our concluding suggestions regarding additional forms of measurement in order to think about
education and redistribution is intended as preliminary and subject to
discussion and revision. We consider that thinking about distribution and
capabilities requires a range of different ways of evaluating social justice in
the provision of education that goes beyond the existing indicators. This is a
key task if we are to be confident of our achievement of EFA by 2015.

Notes

1. This section of the chapter compresses a series of arguments we have made in
   great detail elsewhere. (Brighouse and Unterhalter 2007)

References

Alkire, Sabina. 2003. *Valuing Freedoms Sen's Capability Approach and Poverty
Reduction.* Oxford: Oxford University Press.

Anthias, Floya. 1998. Evaluating Diaspora: Beyond Ethnicity? *Sociology* 32 (3):
557–580.

Arnot, Madeleine, and Diane Reay. 2006. Power, pedagogic voices and pupil talk:
The implications for pupil consultation as a transformative practice. In
Knowledge, power and education reform, edited by R. Moore, M. Arnot, J. Beck,
and H. Daniels. London: Routledge.

Black, R, and H. White, eds. 2003. *Targeting development: Critical perspectives on the
Millennium Development Goals.* London: Taylor and Francis.

Brighouse, Harry. 2002. What rights (if any) do children have? In *The moral and
political status of children,* edited by A. Archard and C. Macleod. Oxford: Oxford
University Press.

Brighouse, Harry, and Adam Swift. 2006. Equality, priority and positional goods.
Ethics 116:471–497.

Brighouse, H. and E. Unterhalter. 2007. Primary goods versus capabilities:
Considering the debate in relation to equalities in education. Paper prepared for
the Philosophy of Education Society of Great Britain Conference, March, Oxford.

Carr-Hill, Roy, Mike Hopkins, Abby Riddell, and John Lintott. 1999. *Monitoring the
performance of educational programmes in developing countries.* London:
Department for International Development.

Craig, Helen, and Ward Heneveld. 1996. *Schools count: World Bank project designs
and the quality of African primary education.* Washington: World Bank. http://
www.worldbank.org/afr/findings/english/find59.htm (accessed March 2003).

Ferdous, A., and A. Rahman. 2000. *Classroom performance in the primary schools in
Bangladesh: Study findings of the Primary School Performance Monitoring
Project.* Dhaka: Primary School Performance Monitoring Project.

Frank, R. 1999. *Luxury fever.* Princeton: Princeton University Press.

Gasper, Des. 2004. Capability approach and subjective well-being. Paper read at the Seminar on Capabilities and Happiness, March, St. Edmund's College, University of Cambridge.

Goldstein, H. 2004. Education for All: The globalization of learning targets. *Comparative Education* 40 (1): 7–14.

Heynemann, S. 1997. Educational quality and the crisis of educational research. In *Education, democracy and development,* edited by R. Ryba. London: Kluwer.

Hirsch, F. 1977. *Social limits to growth.* London: Routledge Kegan Paul.

Levacic, R., and A. Vignoles. 2002. Researching the links between school resources and student outcomes in the UK: A review of issues and evidence. *Education Economics* 10 (3): 313–331.

Lupton, R., C. Brown, F. Castles, and A. Hempel-Jorgensen. 2006. Group, class and school composition effects: Some early results from the Harps Project. Paper read at BERA Annual Conference, September 6–9, Warwick University.

Motala, Shireen. 2003. Review of the financing, resourcing and costs of education in public schools: A commentary. *Quarterly Review of Education and Training in South Africa* 10 (1): 2–12.

Nelson Mandela Foundation. 2005. *Emerging voices: A report on education in South African rural communities.* Pretoria: HSRC Press.

Phillips, Anne. 1999. *Which equalities matter?* Cambridge: Polity Press.

Pogge, Thomas W. 2003. Can the capability approach be justified? *Philosophical Topics* 30 (2): 167–228.

Robeyns, Ingrid. 2006. Measuring gender inequality in functionings and capabilities: Findings from the British Household Panel survey. In *Gender disparity: Manifestations, causes and implications,* edited by P. Bharati and M. Pal. New Delhi: Anmol Publishers.

Saito, Madoka. 2003. Amartya Sen's capability approach to education: A critical exploration. *Journal of Philosophy of Education* 37 (1): 17–33.

Samoff, J. 1994. *Coping with crisis: Austerity, adjustment and human resources.* London: Cassell.

Sen, Amartya. 1985. *Commodities and Capabilities.* Amsterdam: New Holland.

———. 1999. *Development as Freedom.* Oxford: Oxford University Press.

Singh, R. B., T. Karki, R. Carr Hill, and S. J. Lohani. 2002. *Updating indicators for basic and primary education sub sector and for BPEP II.* Kathmandu: Department of Education.

South Africa. 2003. *Report to the Minister: The financing, resourcing and costs of education in public schools.* Pretoria: Department of Education. http://education.pwv. gov.za/Policies%20and%20Reports/2003_reports/reports_2003%20Index.htm (accessed May 2003).

UNESCO. 2000. *Education for All: Status and trends 2000.* Paris: UNESCO. http://unesdoc.unesco.org/images/0011/001198/119823e.pdf.

———. 2003a. *EFA Global monitoring report:Gender and education for all, the leap to equality.* Paris: UNESCO. http://portal.unesco.org/education/en/ev.php-URL_ID=23023&URL_DO=DO_TOPIC&URL_SECTION=201.html.

———. 2003b. *EFA global monitoring report: Gender and education for all, the leap to equality (Summary report)*. Paris: UNESCO. http://www.efareport. unesco.org/ 07.11.03.

———. 2005. *EFA Global monitoring report 2005: The quality imperative*. Paris: UNESCO. http://portal.unesco.org/education/en/ev.php-URL_ID=35874& URL_DO=DO_TOPIC&URL_SECTION=201.html.

Unterhalter, Elaine. 2003. The capabilities approach and gendered education: An examination of South African complexities. *Theory and Research in Education* 1 (1): 7–22.

———. 2005. Global inequality, capabilities, social justice: The millennium development goal for gender equality in education. *International Journal of Educational Development* 25:111–122.

———. 2006. Global inequalities in girls' and women's education: How can we measure progress? Paper read at the International Conference of the Human Development and Capability Association: Freedom and Justice, August 29–September 1, Groningen, the Netherlands.

Unterhalter, Elaine, Rajee Rajagopalan, and Chloe Challender. 2005. *A scorecard on gender equality and girls' education in Asia 1990–2000. Report prepared for UNESCO Bangkok, September 2004.* Bangkok: UNESCO. http://www2. unescobkk.org/elib/publications/gender_equality_asia/gender-equality.pdf.

Unterhalter, Elaine, Jake Ross, and Mahmudul Alam. 2003. A fragile dialogue? Research and primary education policy formation in Bangladesh 1971–2001. *Compare* 33 (1): 85–99.

Watkins, Kevin. 2000. *The Oxfam Education Report*. Oxford: Oxfam.

Wolf, A. 2001. *Does education matter? Myths about education and economic growth.* London: Penguin Books.

World Declaration. 1990. *World conference on Education for All: Meeting basic learning needs. Final Report.* New York: UNESCO Inter-agency Commission.

Young, Marion. 2006. Defining valued learning and capability. Paper read at the International Conference of the Human Development and Capability Association: Freedom and Justice, August 29–September 1, Groningen, the Netherlands.

5

Gender Equality, Education, and the Capability Approach[1]

Elaine Unterhalter

What is gender equality in education? On one level the question is straightforward as this goal has been an aspiration of social policy for decades in many countries. But the simple question raises some difficult issues. Is gender equality in education about equal numbers of boys and girls in different phases of schooling? Is it about boys and girls gaining equivalent levels of examination passes? Is it about gender relations and social practices in schools and how these may be made more equitable? Or is it a combination of these? This chapter critiques different interpretations of gender equality and education to make the argument that our perspectives on what counts as gender equality have real policy and practice effects.

Ramya Subrahmanian, in reviewing approaches to gender equality in education associated with the Millennium Development Goals (MDGs), has written about gender equality as a normative ideal. She distinguishes this from parity, the practice of ensuring equal numbers. She then goes on to utilize a framework associated with discussions of rights, and distinguishes gender equity in access to education, in the provision of education, and in the outcomes of education (Subrahmanian 2005). Her argument highlights how much of the difference between men and women has been naturalized and taken for granted, and how important it is in moving to a notion of substantive equality to recognize that men and women start from different places.

This discussion is a very helpful illumination of some of the overlaps in policy discussion, where gender equity has become a catchall phrase for a range of policies and practices, not all of which are about women's rights and more equitable social relations. But Subrahmanian's distinction between different forms of rights and gender equity begs the question about the content of the normative notion of equality that underpins the discussion. In this chapter I consider some of the critiques of equality as a concept in an attempt to think through some of the ambiguities and difficulties entailed in researching gender equality in education and developing transformative policy. In doing so, I draw on an interpretation of the capability approach and distinguish this way of thinking about equality in education, and thus gender equality, from some of the other major contributions to this field.

The first section looks at different meanings of gender and how these link with contrasting approaches to equality. In the following sections, three different approaches to thinking about equality in education are explored in relation to how they bear on understanding gender equality in education. The three approaches to equality discussed are distributive equality, equality of condition, and equality of capabilities. In concluding the chapter I highlight how the three different approaches may usefully be seen to complement each other in effecting a full meaning of gender equality in education as policy and practice.

Different Meanings of Gender, Education, and Equality

There are at least three different ways that gender equality can be understood. These connect to an analysis with different currents in the literature regarding gender and education.

Gender equality can be understood as if gender was a descriptive characteristic of a person, meaning no more than having short hair or a birthday in September. In many campaigns for women's rights to vote, to work, and to enter education, this argument has been made that gender was not a significant bar to political and economic participation or the acquisition and utilization of knowledge. Thus equality here is concerned with either distributing equal amounts of schooling (say five years) regardless of gender, or respecting all people with regard to the quality of education, paying no heed to "incidental" attributes such as gender. It is this meaning of gender equality that has been most utilized by international education policymaking bodies and that is articulated in the gender parity target of the third MDG. This is the stuff of much survey research in education, assessing numbers of girls and boys at different levels of schooling and in different achievement bands in examinations (for some examples see Unterhalter 2005). Gender equality is interpreted as equal numbers of boys and girls enrolled in school, attending classes, and progressing to complete examinations. The assumption is that if equal numbers are present, or achieving well, nothing further then needs to be done to change schooling for girls or boys.

In sharp contrast to this value-neutral meaning of gender, the concept can also be understood in terms of the ways in which gender is known to structure social relations, language, and history between groups. Here gender is saturated with meanings, and the nature of the social processes that form and reform these relations are highly salient. Gender works to position groups of women and men in particular ways legally, financially, culturally, educationally, and discursively. While some people might be able to effect changes in the structures that shape their lives, these changes are always partial and contingent. Equality here is a much more substantive, but also much more fluid notion concerned with reshaping social relations and institutions and contesting meanings on individual, intersubjective, organizational, and representational levels. In this analysis, equality is struggled over through social practices in relation to specific identities and locations, which include education sites. The methodological challenge is not just counting numbers, but understanding relationships and practices. Much analysis rests on documenting inequalities in power and the gender regimes entailed in these allocations and portrayals.

Because much of the work in education concerned with this understanding of gender has been done by sociologists, who have been interested in documenting forms of power associated with gender regimes and gendered discourses, relatively little attention has been given to considering the meaning of equality. Thus, for example, three recently published substantive reviews of scholarship in the field of gender and education contain extensive analysis of gendered practices in education, but scant consideration of meanings of equality (Skelton and Francis 2003; Skelton, Francis, and Smulyan 2006; Leonard 2006). Indeed, the influence of poststructuralism has helped develop a literature concerned with the politics of difference in schools, documenting differences linked to masculinities, femininities, class or race identities, and questions of sexuality (e.g., Swain 2004; Morrell 1998; Epstein and Johnson 1998; Hey 2006; Youdell 2006). Implicitly, the form of equality being struggled over here is not just equal amounts of schooling, as this is generally taken as given, but some affirmation of equal esteem or equal concern for all within the field of schooling or education (however this is defined). Generally, however, in this scholarship the language of equality or equity is not deployed.

There is a third meaning of gender that has affinities with the two I have outlined, but is, in my view, distinctive. In this meaning, gender signals something about the attributes of the person while acknowledging that these are changeable and entail freedom and agency as well as the constraints of constructed social relations, for example, girls from different families, cities, or historical contexts might "do gender" differently at different moments in their own lives, sometimes because of the demands of social structures, but sometimes because of negotiations and contestations. Some of the discussions in comparative and international education regarding gender highlight this, showing, for example, how women attending literacy projects in Nepal "do gender" differently in settings organized by international development assistance and in settings that link to community or personal priorities (Robinson-Pant 2000). Other work highlights different understandings in global and local terrains, for example, the understandings of women's empowerment and education amongst indigenous groups in the Peruvian Amazon are very different from those of a largely urban, global women's movement (Aikman 1999). However, this third meaning is not the same as the post-structuralist concern with difference outlined above, as issues of poverty and global inequality frame the discussion. This third meaning thus resonates with an acknowledgment of the power of global social relations and the intersections of gender, race, and class across national boundaries, but suggests these might be challenged and shifted. This third sense of "gender" places emphasis on changing gender identities and on the way gender changes the amounts or form of education that people may need in a context of shifting global relations. Accounts of women struggling not only with gender hierarchies within education institutions but also with received gender identities and global politics so that they can access education illuminate this (Unterhalter and Dutt 2001; Vavrus 2005; Kirk 2004). It is this sense of gender that I think resonates with Amartya Sen's capability approach, as I discuss below. In this approach gender is both a feature of the constraints on a capability set, and also part of the negotiations and agency entailed by the notion of capability.

These three different meanings of gender signal different approaches to thinking about equality and justice. Nancy Fraser (1997, 2005) has highlighted the importance of distinguishing redistributive, recognitional, and representational dimensions of justice, drawing on race and gender to illustrate this. Thus, in thinking about redistributive justice, concerns with gender must disappear so that gender does not color equality of treatment, for example, whether a girl or a boy is given a textbook to take home to complete a homework assignment or the amount of time a teacher spends explaining a difficult idea. Redistributive ideas, in Fraser's analysis, work with the first meaning of gender I identified, and equality entails equal amounts. Recognitional approaches to justice highlight the significance of gender in stressing how girls may have different learning styles, how women teachers work within particular gender regimes, and how the politics of the family entail negotiations between mothers and fathers regarding children's schooling. Gender here resonates with the second meaning I identified, while equality, understood in terms of equal regard, is inflected through an appreciation of considerable difference. Madeleine Arnot has charted how the history of feminist struggles in education in the UK has emphasized either redistributional or recognitional poles (Arnot 2006). Fraser's representational dimension of equality is concerned with whether women participate on terms of parity with men in decision making and whether the languages and routes to power they have are equal. The representational dimension of equality, which is concerned with aspects of process equality, and acknowledges that different procedures might be required for different groups (Young 2000, 2001; Phillips 2004; Young 1990) has, as yet, not been examined empirically across the terrain of global social division that I associated with the third form of thinking about gender.

These contrasting ideas about gender and approaches to thinking about equality entail different ways of thinking about equality in education.

Conceptualizing Equality in Education: Distributive Equality

One approach to thinking about equality in education draws on the Anglo-American egalitarian literature in political philosophy. In considering whether this is helpful for thinking about gender equality we need to acknowledge how this literature works with ideas in economics that talk about equalities in income or resources, suggesting that in applying this to gender and education we are looking at equal amounts of education or certificates of equal value. A second approach found in the work of Kathleen Lynch and John Baker (2005) draws out how equalities in education are associated with equal empowerment and enabling conditions for people to pursue a good life. As explained below, this is a thicker notion of equality than recognition and equality of esteem, but works well with the second meaning of gender I identified. A third approach, which stresses equality of capabilities, is close to equality of condition but considers aspects of changing social arrangements that connect education, health, and economic and social policy, and thus has particular resonance with regard to problems of intersecting global inequalities. Thus gender equality in education is made between spheres of social policy and not only within them.

Ingrid Robeyns has argued that the egalitarian literature dealing with inequality in welfare economics, and the inequality literature in analytical political philosophy are of rather limited use to look at the question what gender equality means (Robeyns 2002). The methodological assumptions in both tend to underplay what is salient in thinking about gender.

They focus on ideal theory and underplay the web of social relations in which gender is enmeshed. Their analysis also often sets up static problems of distribution, ignoring the way in which individuals do not just have equal amounts but are relationally engaged in making meanings and taking action about those amounts. Robeyns concludes that "both literatures are not only normatively but also methodologically individualistic. They look at individuals rather than groups. The social processes/structures that make up "gender" (in its thick understanding) thus have no easy place in this analysis" (Robeyns 2007). Much of their work looks at individual- or household-level income inequality, which reveals very little about quality of life and equality in education. Thus this literature compares individuals, as though they were all potentially equal and outside history. It does not compare groups or take on board that the history of some groups is particularly salient for the question of equality. Group inequalities, such as gender or race inequality, often persist in education regardless of the effort particular individuals may exercise, for example, to learn hard at school or work hard as a teacher. Similar points are made by Iris Young (2000), and Avigail Eisenberg has distilled how her analysis can be applied in education, showing how oppression, domination, and the silencing and exclusion of difference require redress, not distributive equality (Eisenberg 2006).

Robeyns's and Young's critiques present some powerful cautions not to seek an answer to what gender equality in education is wholly through the literature on distribution. The significance of their critiques is evident if we look at gender equality in education through the lens of a particular discussion between egalitarians regarding equality or priority (Parfit 2002). Parfit distinguishes between strict egalitarians who claim that there are ethical objections to giving some groups more than others and prioritarians who argue that in order to effect improvements in the conditions of the least advantaged some inequalities are necessary. In order to consider the explanatory reach of this and contrasting views on equalities in education, I want to look at each in relation to issues that have arisen for me in the course work in a team looking at the distribution of instructional materials to all primary schools in Kenya as part of the government's introduction of free primary education (FPE) supported by international development assistance.[2]

When the government of Kenya introduced FPE in 2003, abolishing school fees and charges for textbooks, it had a clear goal to expand education provision and achieve gender equality in basic education, ensuring that all girls and boys were supported to complete eight years of basic schooling. The government's aim has been that schooling will be of equal quality regardless of the levels of poverty of families, schools, or districts. This is a strict egalitarian position. However, the government also faces problems in that generally girls do less well in mathematics and science than boys in the Kenyan Certificate of Primary Education (KCPE) that certifies the end of this phase of schooling (Elimu Yetu 2005). Girls and boys in areas with somewhat lower levels of poverty do the worst, possibly because in these areas richer children are in private schools. The view of the prioritarians is that in order to improve the learning and teaching in mathematics and science at school for the girls who do worst in these subjects in the KCPE, that is, the worst off, the girls should get priority in any policies regarding the distribution of education.

There are many ways the government might empirically decide to support the worst off. Here are two hypothetical possibilities. The ministry of education might decide to develop a cadre of highly specialized and well-paid mathematics and science teachers from the best graduates, the majority of whom are men, generally from more affluent families as they have passed mathematics and science at a good enough level to secure university entry. These very able students, it could be argued, are best placed to develop an in-depth understanding of how to help the poorest children overcome obstacles to learning about mathematics and science; the difficulties of their work with these children will require higher levels of pay to incentivize them to make this commitment to supporting education. On the other hand, the ministry of education might argue that evidence suggests that it is women teachers, who have overcome difficulties in learning mathematics and science themselves, who are best suited to teach the poorest girls. They will appreciate their circumstances. Therefore, women with lower general scores in mathematics and science at the end of secondary school should be selected in place of men for places on teaching degrees. Such a policy of increased quotas for women teachers is a feature of a number of countries (Raynor and Unterhalter 2007 forthcoming; UNESCO PROAP 2000). At issue here is that in order to achieve some future level of gender equality in education, here understood as equal amount of education attainment in mathematics and science, some gender and education inequalities are necessary in terms of who is admitted to teaching degrees.

Thus the introduction of some gender inequalities may lead to a situation that is considered more just in a prioritarian view than the existence of no inequalities at all, if this means that the worst off are better off in a situation with inequalities than in a situation with strict equality. Thus, we could compare a society where men and women from richer and poorer backgrounds had roughly equal amounts of education as measured by years in school, but the overall levels of education were not that high, with the possible scenario outlined above making it better for some—men from elite schools who gain access to highly paid jobs as specialist mathematics and science education advisers—and worse for no one as the poorest girls and boys will have some better teaching in mathematics and science. If we

Outcomes in Districts A and B, by gender and wealth

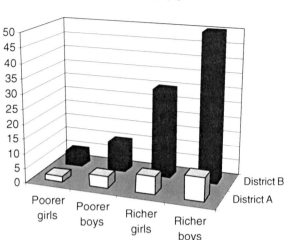

Figure 5.1   Contrasting distribution of young men and women completing university degrees in mathematics and science in two districts

look at this diagrammatically (figure 5.1), the numbers represent the proportion of children from each class background in a particular district completing university courses in science and mathematics degrees, which ensure highly paid jobs.

In district A, only a small proportion of children from rich and poor backgrounds gain access to this level of qualification, and although there are gender and class differences, they are not that large. In district B, the gender and class differentials have increased but no group is worse off than it was in district A. Thus district B is Pareto superior[3] as equalizing to make district B more like district A would make district B worse off.

Pareto optimality is widely valued. But the idea that the prioritarian position licenses these inequalities is subject to challenge (Cohen 1995). Drawing on Cohen's critique, the bald statement of the prioritarian argument is that all the children in districts A and B had education of equal quality and all had equal opportunities to participate in discussion about how school and higher education connected and what the opportunities in the workplace as a result of studying mathematics and science would be. The outcome represented by the proportion of boys and girls entering university is thus assumed not to be the result of histories of discrimination and exclusion in relation to quality of teaching, opportunities for learning, and levels of access to information about work, but the outcome of whether or not the different children worked hard and passed examinations at the appropriate level. Much of the literature on gender, class, and access to higher education shows that this is not the case. There is no level playing field. Thus the first egalitarian response to the prioritarian question is that equality of outcomes cannot be disassociated from equality of opportunities that take the deep and long-lasting effects of gender and poverty on board. Gender equality is thus about a link between opportunities and outcomes and it entails processes for inclusion in decision making as a core component of equality.

Further, Cohen's general discussion suggests that there might be other scenarios than the contrasts drawn between districts A and B. In district C, which, let us suggest for the sake of argument, has a population that is the same size as that of district B, the chief education officer has put in place policies to distribute resources between schools so that more science textbooks and specially trained teachers are located in schools attended by the poorest families. In addition, intensive programs reviewing gender equality and poverty eradication are undertaken ensuring women gain ownership of land, laws prohibiting gender-based violence and early marriage are enforced, and the citizen rights of the poorest to own property and access social benefits are respected. In district C, the numbers of children completing a science degree is distributed as indicated in figure 5.2.

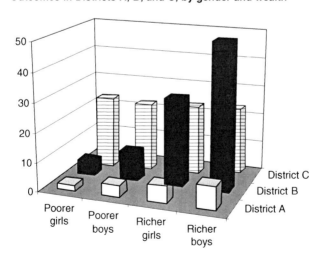

Figure 5.2 Contrasting distribution of young men and women completing university degrees in mathematics and science in three districts

A total of 96 children from districts A and B complete science degrees, but in district C the effects of the antipoverty and gender equality programs in schools have been the virtual disappearance of inequalities on the basis of gender and poverty. One of the questions that Cohen's critique needs us to ask is whether the scenario in district C is feasible. What are the conditions in which richer girls and boys might be willing to study hard even though they know that their chances of succeeding in prestigious science degrees are not necessarily ensured because of the redistribution to children from poorer families? We can expand this by considering whether gender equality must also entail class equality as a number of other scenarios could be sketched where one or the other was achieved but not both.

What this discussion illuminates is that some of the distributional questions concerning gender equality in education are not as simple as the assumption about equal amounts of schooling suggested. A version of the prioritarian solution might be appropriate to some short-term interventions in societies marked by poverty and gender inequalities. But this form of distributive approach to gender has considerable limitations as it considers education credentials to work like forms of income. It thus pays scant attention to how equality is lived and produced inside schools. It is here that the second approach to thinking about equality is particularly useful.

Equality in Education: Equality of Condition

Lynch and Baker (2005), by contrast, argue for a very thick notion of equalities in education. They point out how social-class inequalities in education derive from economically generated inequality and how education institutions can promote these through approaches to admission, grouping children on the basis of perceptions about ability, and the bias in systems of curriculum design and assessment in favor of language skills and mathematics. They suggest an equality in education that develops respect and recognition for diverse children, the need for valuing care, love, and solidarity as much as narrow forms of academic achievement, and equality in power to democratize planning and discussion. Gender equality using this framing would intersect with concern with developing other forms of enabling condition and empowerment for all children and all those working in education. In this analysis it is not outcomes or achievements that are equalized but conditions in school. The prioritarian trade-offs have little explanatory power because each education setting has a complex array of social arrangements in which conditions will need to be equalized. The arguments made for education that values difference and seeks to address inequalities of power, as developed by Eisenberg

(2006) from the elucidation of Young's (2000; 2001) work in relation to education, has many affinities with this position.

However, if we return to the questions of policy in relation to poverty, gender, and the KCPE results outlined above, there are a number of unresolved issues. In this second analysis, every school requires a maximum amount of resources, which can be understood as money; knowledge about how to put ideas about enabling education into practice; support from children, parents, communities, and state officials. Every school matters and every form of inequality matters too. The vision of equality is substantive and while enormously important as a goal to be aimed at, it might be too overwhelming a policy challenge for governments, their international development partners, and civil society activists to effect.

Equalities in Education: The Capability Approach

A third view emerges from thinking about gender equality in education in relation to the capability approach. Amartya Sen takes issue with the approaches to evaluating social policy that focus on the aggregated benefits an initiative has for the whole society or for future generations, without regard to how it affects individuals. According to these views, for example, investing in education for women and girls is justified by its benefits not for them, but for the societies they live in. These approaches to evaluation do not look at whether any adult or child has been discriminated against in the provision of education, because the education is not for those individuals but for a larger grouping—the community, then nation, and future generations. These collectivities might be weakly interested in gender equality in education or might have a formal commitment to it, but when practices or processes are assessed, systems or organizations are reviewed and not aspects of the flourishing of particular individuals.

Sen's alternative is concerned with evaluating social policy, including education, without ignoring individual aspirations or dictating social benchmarks. He argues that we must evaluate policy in the space of capabilities, which he defines as valued "beings and doings." The foundational idea is that what matters to people is that they are able to achieve actual *functionings*, that is, "the actual living that people manage to achieve" (Sen 1992, p. 52). In this there is considerable overlap with the notion of equality of condition. But importantly, the idea also assumes that social conditions in, for example, health and housing, and access to the labor market will support gender equality.

Sen's work on capabilities has developed out of a critical engagement with welfare economics. He conceptualized capabilities partly in response to a general pessimism that welfare economics could not deliver useful forms of interpersonal comparison (Sen 2002, pp. 70–71 and 78–79). The implication of what was seen as the weak informational base of welfare economics entailed that it was impossible to develop policy on schooling, for example, as one could not know how schooling was valued by all individuals in a particular society; how schooling was ranked relative to other important goods, for example, that girls remain at home and help with housework rather than go to school; or whether aspirations for or against schooling were the "real" views of an individual, or the views they had to adopt because of powerful customs that dictated appropriate behavior. Sen's capability approach sought to answer the question regarding what kind of comparisons could be made regarding quality of life between people and what kind of information could provide for those comparisons.

The kind of answer he provides is best illustrated through his contribution to debates on social justice. A broad range of writers concerned with social justice have argued over whether equality or social justice or the fulfillment of human rights between people should be assessed in terms of resources or welfare (Clayton and Williams 2002, pp. 8–13).

Arguments claiming that what is to be compared is access to resources, point out that the extent to which individuals have equal command over resources, for example, actual places in school and the actual rights and capacity to take up those places, is the most appropriate measurement (Dworkin 2000). Thus, for example, in Kenya, to what extent do children whose families live by nomadic pastoralism have command over the resources the government supplies for schooling when this is only located in settled communities?

The welfarist argument notes that what really matters to people is satisfaction, and that people convert resources into satisfaction at different rates. So what we should focus on is welfare, understood variously as happiness, pleasure, or preference-satisfaction. The most influential version of the welfarist view in political philosophy does not focus directly on preference satisfaction but, instead, on opportunities for welfare, saying that inequalities of welfare are justified if and only if they reflect responsible choices against a background of equal opportunity (Arneson 2002). Thus, for example, two Kenyan schoolgirls might both wish to complete secondary school. School for one is a short distance from her home and has been long established with well-trained teachers. School for the second is 100 km from the area where her nomadic pastoralist family makes a living. Satisfying her preferences for completing secondary school requires the expenses incurred in traveling long distances to school, boarding facilities, and pastoral care because of the long periods of separation from her family. In addition, education officials will need to be employed to mobilize support from her community that is opposed to this level of schooling for girls. Arneson's argument is that these inequalities in government expenditure of resources are justified only if she and her family come to the decision for her to continue in school against a background of fair equality of opportunity, that is, all girls will have access to secondary school and ability to access jobs outside their homes after completing school. Welfarism, however, is widely criticized for its inability to deal appropriately with the problem of expensive tastes, for example, the possibility that the girl from northern Kenya would only be satisfied with schooling at an elite boarding school in South Africa, not the nearest school to where her family lived. A second criticism is that writings on welfare tend to treat expensive preferences as no less urgent than disabilities (Brighouse 1996; Schaller 1997). One could make a similar riposte that the problem of expensive tastes is given considerable airing in this literature when the urgent needs of nearly 1 billion adults and children with little or no schooling, two-thirds of whom are women and girls (Unterhalter 2007) are not written about.

Sen, then, shifts the argument by saying we need to look between resources and achievements, and make interpersonal comparisons in the "space of capabilities," that is, we need to look at the freedom people have to formulate capabilities, "valued doings and beings," and thus convert resources into functionings they value. Thus agency and freedom to make up one's mind about schooling as a valued end and convert one's aspirations regarding schooling into valued achievements lie at the heart of Sen's capability approach and distinguish it from other positions. The centrality of freedoms and the ways in which freedoms in education connect with other freedoms mark the approach out from the more general equality of condition argument discussed above. While Sen argues that such decisions must be taken by each individual, how a society comes to decide on what are the valued capabilities and what resources should sustain this is a matter for each society, and there may be considerable variation between societies and between different levels of governance in any particular society. In this the approach talks to the third comparative view of the meaning of gender outlined earlier. A conception of gender made and remade socially and individually is evident.

An issue the capability approach has to confront is what happens when individuals in articulating "valued beings and doings" merely reflect what is expected of them, for example, women who do not claim adequate education for themselves or their daughters because they believe that it is appropriate for women not to be educated or only educated to a very low level. The capability approach presents a partial answer to this by suggesting that the more valued states are those that expand capabilities and that a very strenuous process of public debate and discussion should be conducted with wide participation so that the nature of what expanded capabilities are is openly determined. This might entail an examination of the ways in which gendered institutions prevent women converting capabilities into functionings. Thus the capability approach highlights the importance of public scrutiny and discussion using a wide range of information as a counterbalance to the difficulty that limits an aspiration might be developed and sustained in private settings.

But what if women have access to the same capabilities, but are less ambitious or do not value education as much as men do, and therefore end up spending less time in school or higher education? The capability approach would suggest this is not just an issue of individual views of desire satisfaction. Implicit in the approach, with its stress on positive freedoms, is the notion that women and men should have the same effective freedom to education—that is, not only formally and legally, but also in terms of being liberated from other constraints such as being forced to do excessive amounts of domestic labor or to care for siblings. Assume that boys and girls or men and women do have the same effective freedom or capability to education, but that girls are told by their parents or wider community that there is no need for them to go to school, either because they will be married at a young age or because education for girls and women is not valued or is seen as a drain on a household's resources. In that case, social customs and the prevailing ethos shrink girls' capability for education; hence the real or effective freedom is reduced to an empty opportunity. In addition, girls themselves may not value education. In these cases, the capability approach suggests an assessment needs to be made with regard to how just these norms and values are that influence girls' preferences, ambitions, and aspirations. In any particular instance, these norms and cultural values would need to be debated and critically scrutinized. Sen does not advocate capabilities that we value as such, but capabilities that we have reason to value, that is, capabilities we value after self-reflection and open debate. Thus, the capability approach is more than simply a proposal to focus on people's capabilities; it also entails a critical engagement with all social, cultural, and other factors that shape people's preferences, expectations, and perceptions, and thus influence which choices are made from the freedoms that we have. It is not only about equality of condition in education but also about equality of conditions and capabilities that bear on education.

A number of commentators have expanded Sen's discussion elaborating the ways in which such self-reflection and public debate might take place (Alkire 2002, p. 43; Robeyns 2003, p. 11). Alkire and Robeyns offer different approaches. Robeyns suggests a number of criteria that can shape public debate on how relevant capabilities can be selected in any particular setting.

These entail public discussion both about capabilities and about how capabilities are to be discussed; sensitivity to the context in which the capability approach is to be applied, for example, local settings with complex gender histories regarding gender equality in education; an approach to thinking about the list both in the ideal, that is, the longed-for far-away future, and in the soon-to-be-implemented future; and that the list of capabilities should include all that are relevant. Thus it would be insufficient to include only limited education when capabilities regarding health, safety, good social relations, and respect might be very necessary to sustain participation in education (Robeyns 2003, pp. 15–18). Robeyns's expansion of Sen thus entails some principles for regulating debate with regard to the selection of capabilities, thus ensuring that aspects of low evaluation of needs are appropriately counterbalanced.

Alkire chooses to supplement Sen's notion of capabilities with ideas about practical reason, that is, a form of deep, moral reflection on the reasons for action and the dimensions of value for a group (Alkire, 2002, pp. 43–59). In Alkire's view, utilizing practical reason will allow for an articulation of capabilities and public consultation will allow assessments of how expansions of capabilities are to be articulated and broadened. Alkire's answer to the question of who defines needs differs from Robeyns's in its conviction regarding the importance of practical reason to any discussion of expanding capabilities. Implicit in her view is an idea that self-harm would not be a reason for action and that the use of practical reason would be a defense against women choosing badly for themselves or their daughters. The possibility that they might however is counterbalanced with ideas about wide public scrutiny, similar in intent, if not in form, to those suggested by Robeyns.

Thus, the capability approach, as developed by Sen and others, would seek to consider gender equality in education in three mutually reinforcing ways: an expansion of capabilities, an expansion of the ways in which capabilities are discussed, and the fora in which those discussions take place. It is thus an extremely demanding approach, but one which talks to Sen's concern with human rights entailing not only the fulfillment of needs, such as education, but also attention to the ways in which needs are formulated and the respect we owe and give each other in that process (Sen 1994, p. 38), a respect which entails consideration of gender equality.

Sen argues that in developing and assessing social policy we must look at individual formulations of well-being and the social arrangements for this while allowing for a range of other complex differences between individuals. He also highlights the need for using a range of different forms of information gathering to ascertain views regarding capabilities (Sen, 2002, pp. 84–86). Thus thinking about gender equality in education in relation to the capability approach suggests we are concerned not just with some predetermined form of education (a resource-based view) and standardized measurement of gender equality, possibly linked to access or achievement, but rather with the nature of education valued by individual women and men and the conditions that allow them to express these views and realize these valued "beings and doings." It, therefore, seems to combine the distributive version of equality and the equality of condition approach outlined above.

Governments using the capability approach, therefore, have an obligation to establish and sustain the conditions for each and every individual, irrespective of gender, ethnicity, race, or regional location to achieve valued outcomes. These may entail acquiring a certain level of educational attainment, but they undoubtedly entail ensuring the freedoms that allow valued outcomes to be articulated and achieved. Thus, for example, failure to ensure conditions where sexual violence in and on the way to school can be identified and checked, would be a failure to ensure freedom for valued outcomes. Similarly, failure to ensure opportunities for a particular group to participate in decision making about valued outcomes would also be a limitation on freedoms or capabilities. Although Sen's capability approach highlights the importance of diverse social settings where capabilities will be articulated, it emphasizes the importance of free forms of discussion and association in articulating capabilities. Sen writes about development as freedom because the freedom to think, talk, and act concerning what one values is a meaning of development that is closer to concern with human flourishing than are narrower notions of a certain level of GDP per capita or a prespecified level of resource provision.

Gender Inequality, Education, and Capability Deprivation

A number of commentators concerned with gender equality point out that Sen's capability approach does not provide a set of prescriptions; it only sets out a general framework, not a substantive theory (Nussbaum 2000, p. 70; Alkire 2002, pp. 28–30; Robeyns 2003, pp. 12–13). These writers have all, in different ways, tried to point out how the capability approach might be put into operation either by using it normatively to specify central capabilities that inform constitution making or in work with small and medium-sized projects defining principles through which relevant capabilities can be selected for evaluation. All these commentators share Sen's key concerns with individuals, agency, and freedom. They also acknowledge the complexity of gender inequality and the ways that this is both located in the forms of social arrangements, such as the hierarchies relating to decision making in education, and in the ways in which groups, such as women and girls in many contexts, who have experienced many centuries of discrimination and subordination, have a very diminished sense of agency.

The capability approach in education requires us to think about the gendered constraints on functionings and freedoms in educational organizations, such as schools or adult literacy classes. It also draws attention to how sometimes, despite relatively high levels of education for girls and women, the legal system, the forms of political participation and economic ownership, or employment and leisure practices limit agency and "substantive" freedom of girls and women, thus entailing capability deprivation. Thus, it is not only substantive equalities in education that must be developed but also those in the social policy areas that bear on education. In some societies, a good proportion of women do complete 12 years of schooling, but then encounter prohibitions on inheritance and property ownership, discrimination in relation to employment, and assumptions about the food they will eat and how leisure time will be used. These arrangements are sometimes normalized, confirmed, or barely challenged by the form and content of what is learned at school.

We can apply the capability approach to thinking about education in a society such as Kenya, marked by poverty and regional and gender inequalities. International development assistance has helped support FPE, but there are sharp differences between partners. Here we might look at education policy and practice at the international, national, regional, or local level. We might also ask whether long-term valued goals concerning education for girls and boys, women, and men have been articulated. Was there a process of consultation open to all (irrespective of class, race, ethnicity, or gender)? Was this well facilitated to bring out the views of individuals who are often silenced and who do not have the skills to articulate their views? Second, we would need to consider whether particular steps (including appropriate resources) to achieve those goals have been put in place and whether these steps are adequate for all, given the long history of inequalities. It might be that achieving certain functionings requires more resources for girls and women than for boys and men. A process of public consultation could clarify why this was so and how some aspects of inequality allowed the whole society to achieve goals it viewed as valuable. Third, we would need to see whether strategies are in place that will safeguard participants in the education initiative from, for example, intimidation, violence, withdrawal of resources by hostile groups at the local level or from groups at a higher level, say, regional, national, international. This might be a very complex process as some violence is family based while some proceeds through use of long-established hierarchies, where forms of obedience or acquiescence are deeply entrenched.

Yet, again, the capability approach alerts us to using public scrutiny and discussion to consider how this aspect of freedom linked to capabilities might be enhanced. Lastly, the policy analyst would need to look at ways in which the articulation of goals in this education initiative mesh with processes to articulate wider goals of what is valued for all in the society. Again this might be a very difficult process, but once again it is through a process of public debate that decisions on this can be reached.

Conclusion

The three approaches to thinking about gender outlined align with the different ways identified for thinking about equalities in education. When confronting the complexities of poverty and gender inequalities in education in a country such as Kenya, where global relationships associated with aid, trade, and education practice are of enormous significance, it seems that a combination of all three approaches is required. Distributive equality links with approaches to thinking about gender descriptively, although the prioritarian problem requires a much thicker notion of gender and social relations. Equality of condition talks to views of gender deeply enmeshed in the relationships and meanings made by school, but suggests few policy and practice pointers to linking education with other areas of social policy. Equality of capabilities considers gender in a range of different settings and is particularly appropriate when thinking of gender across different historical contexts, different formations of globalized social relations, and the connections between education and health, food, politics, or employment. But this approach to equality draws on appreciations of gender and diversity that are not given at the outset. Gender inequalities in education cannot be fully addressed by any single approach to gender or equality. The complexity and the import of social justice initiatives suggest all three are needed to complement each other and thus enhance policy and practice.

Notes

1. This chapter builds on two earlier papers: "Education, Capabilities and Social Justice" commissioned by UNESCO in 2003 as a background paper for the Global Monitoring Report (Unterhalter 2003) and "Gender Equality and Education in South Africa: Measurements, Scores and Strategies" presented at a symposium on "Gender Equity in Education" organized by the British Council and the HSRC in Cape Town in June 2004 (Unterhalter 2004). My thanks to Chris Colclough and Ingrid Robeyns for critical feedback on the UNESCO paper, to participants in the Cape Town symposium for useful comments, and to Melanie Walker, Ingrid Robeyns, and Harry Brighouse for very helpful

suggestions on this revision. This work has developed in the context of ongoing association with the *Beyond Access* project, and the contribution of Sheila Aikman, Chloe Challender, Amy North, and Rajee Rajagopalan over many years to my thinking on these issues is gratefully acknowledged.

2. The examples given in this chapter arise from ideas that presented themselves to me in the course of my work with the study team in 2006. They are not a record of the findings of that team. The full report of the team (Yates et al. 2007, forthcoming) is in press at the time of writing.

3. One state or district (A) is strongly Pareto superior to another (B) when everyone is better off in state A than in state B. A may be weakly Pareto superior to B if at least one person is better off in A and no one is worse off.

References

Aikman, Sheila. 1999. Schooling and development:Eroding Amazon women's knowledge and diversity. In *Gender, Education and Development: Beyond access to empowerment,* edited by C. Heward and S. Bunwaree. London: Zed.

Alkire, Sabina. 2002. *Valuing freedoms: Sen's capability approach and poverty reduction.* Oxford: Oxford University Press.

Arneson, R. 2002. Liberalism, distributive subjectivism and equal opportunity for welfare. In *The ideal of equality,* edited by M. Clayton and A. Williams. Basingstoke: Palgrave Macmillan.

Arnot, Madeleine. 2006. Gender equality, pedagogy and citizenship: Affirmative and transformative approaches in the UK. *Theory and Research in Education* 4 (2): 131–150.

Brighouse, Harry. 1996. Egalitarianism and equal availability of political influence. *Journal of Political Philosophy* 4 (2): 118–141.

Clayton, M., and A. Williams. 2002. Some questions for egalitarians. In *The ideal of equality,* edited by M. Clayton and A. Williams. Basingstoke: Palgrave Macmillan.

Cohen, G. A. 1995. The Pareto argument for inequality. *Social Philosophy and Policy* 12:160–175.

Dworkin, R. 2000. *Sovereign virtue.* Cambridge: Harvard University Press.

Eisenberg, Avigail 2006. Education and the politics of difference: Iris Young and the politics of education. *Educational Philosophy and Theory* 38 (1): 7–23.

Elimu Yetu. 2005. The challenge of educating girls in Kenya. In *Beyond Access: Transforming policy and practice for gender equality in education,* edited by S. Aikman and E. Unterhalter. Oxford: Oxfam.

Epstein, D., and R. Johnson. 1998. *Schooling sexualities.* Buckingham: Open University Press.

Fraser, Nancy. 1997. *Justice Interruptus.* London: Routledge.

———. 2005. Reframing justice in a globalising world. *New Left Review* 36:69–88.

Hey, Valerie. 2006. Getting over it. Reflections on the melancholia of reclassified identities. *Gender and Education* 18 (3): 295–308.

Kirk, Jackie. 2004. Promoting a gender just peace: The roles of women teachers in peacebuilding and reconstruction. *Gender and Development* 2 (3): 50–59.

Leonard, Diana. 2006. Gender, change and education. In *Handbook of gender and women's studies,* edited by K. Davis, M. S. Evans, and J. Lorber. London: Sage.

Lynch, Kathleen, and John Baker. 2005. Equality in education: An equality of condition perspective. *Theory and Research in Education* 3 (2): 131–164.

Morrell, Robert. 1998. Of boys and men: Masculinity and gender in Southern African studies. *Journal of South African Studies* 24 (4): 605–630.

Nussbaum, Martha C. 2000. *Women and human development: The capabilities approach.* Cambridge: Cambridge University Press.

Parfit, D. 2002. Equality or priority? In *The ideal of equality,* edited by M. Clayton and A. Williams. Basingstoke: Palgrave.

Phillips, Anne. 2004. Defending equality of outcome. *Journal of Political Philosophy* 12 (1): 1.

Raynor, Janet, and Elaine Unterhalter. Forthcoming 2007. Promoting empowerment? Contrasting perspectives on a programme to employ women teachers in Bangladesh, 1996–2005. In *Women teaching in South Asia: An edited collection,* edited by J. Kirk. New Delhi: Sage.

Robeyns, Ingrid. 2002. Gender inequality: A capability perspective. Unpublished PhD thesis, University of Cambridge, Cambridge.

———. 2003. Sen's capability approach and gender inequality: Selecting relevant capabilities. *Feminist Economics* 9 (2–3): 61–91.

———. 2007 forthcoming. Sen's capability approach and feminist concerns. In *The Capability Approach: Concepts, measures and applications,* edited by S. Alkire, F. Comim, and M. Qizilbash. Cambridge: Cambridge University Press.

Robinson-Pant, A. 2000. Women and literacy: A Nepal perspective. *International Journal of Educational Development* 20 (4): 349–464.

Schaller, W. 1997. Expensive preferences and the priority of right: A critique of welfare-egalitarianism. *Journal of Political Philosophy* 5 (3): 254–273.

Sen, Amartya. 1992. *Inequality re-examined.* Oxford: Oxford University Press.

———. 1994. Freedom and needs. *New Republic,* January 10–17: 31–38.

———. 2002. *Rationality and freedom.* Cambridge, MA: Harvard University Press.

Skelton, C. B., and B. Francis, eds. 2003. *Boys and girls in the primary classroom.* Buckingham: Open University Press.

Skelton, C. B., B. Francis, and L. Smulyan. 2006. *Handbook of gender and education.* London: Sage.

Subrahmanian, Ramya. 2005. Gender equality in education: Definitions and measurements. *International Journal of Educational Development* 25:395–407.

Swain, Jon. 2004. Sharing the same world: Boys' relations with girls during their last year of primary school. *Gender and Education* 17 (1): 75–91.

UNESCO PROAP. 2000. *Increasing the number of women teachers in rural schools: A synthesis of country case studies; South Asia.* Bangkok: UNESCO Principal Office for Asia and the Pacific.

Unterhalter, Elaine. 2003. Education, capabilities and social justice. Background paper prepared for UNESCO Global Monitoring Report. http://portal.unesco.org/education/file_download.php/Education%2C+capabilities+and+social+justice.. doc?URL_ID=25755&filename=10739918061Education%2C_capabilities_

and_social_justice..doc&filetype=application%2Fmsword&filesize=67072&
name=Education%2C+capabilities+and+social+justice..doc&location=user-S/
(accessed July 2005).

———. 2004. Gender equality and education in South Africa: Measurements,
scores and strategies. In *Gender equality in South African Education 1994–2004*,
edited by L. Chisholm and J. September. Capetown Human Sciences Research
Council (HSRC).

———. 2005. Fragmented frameworks: Researching women, gender, education
and development. In *Beyond Access: Developing gender equality in education*,
edited by S. Aikman and E. Unterhalter. Oxford: Oxfam.

———. 2007. *Gender, schooling and global social justice*. Abingdon/New York:
Routledge.

Unterhalter, Elaine, and Shushmita Dutt. 2001. Gender, education and women's
power: Indian state and civil society intersections in DPEP (District Primary
Education Programme) and Mahila Samakhya. *Compare* 31 (1): 57–73.

Vavrus, F. 2005. Adjusting inequality: Education and structural adjustment policies
in Tanzania. *Harvard Educational Review* 75 (2): 174–201.

Yates, C., J. Limozi, Ole Kingi, Moses W. Ngware, Jacinta Ndambuki, Johnathan
Casely, Lawrence Barasa, et al. 2007, forthcoming. *Delivering quality and
improving access: An impact evaluation of the instructional materials and inser-
vice teacher training programmes*. Nairobi: Report prepared for the Ministry of
Education.

Youdell, Deborah. 2006. *Impossible bodies, impossible selves: Exclusions and student
subjectivities*. New York: Springer.

Young, Iris Marion. 1990. *Justice and the politics of difference*. Princeton: Princeton
University Press.

———. 2000. *Inclusion and Democracy*. Oxford: Oxford University Press.

———. 2001. Equality of whom? Social groups and judgements of injustice.
Journal of Political Philosophy 9 (1): 1–18.

6

Measuring Capabilities: An Example from Girls' Schooling

Rosie Vaughan

This chapter is primarily concerned with how the capability approach might be used to explore outcomes to girls' education from a rights perspective, and subsequently how existing educational measurements might fit into this framework. In the first section, I briefly review a number of debates surrounding educational measurement. In the second section, I outline how we might theorize the relation between capabilities and educational inequalities, distinguishing between education as a functioning in itself, and education as a facilitator of other functionings. In the last section, with specific reference to gender concerns, I explore how the capability approach could be employed to investigate further how different capabilities are gained from different types of formal education, with particular attention to the confirmation or elimination of gender stereotypes through education.

Girls' education has become one of the most prominent features of development discourse over the past 20 years. Yet, despite an overall rise, global levels of girls' enrollments have continued to lag significantly behind those of boys; further, the concern for equality also extends to progress through school and to the distribution of learning outcomes. While rights-oriented theorists have argued for change on the grounds of equal entitlements, research has also suggested strong links between increasing levels of female education and economic growth, through decreasing fertility, decreasing mortality, and increasing productivity. Calls for greater attention to "gender equality" have subsequently come from a diverse group of voices, from rights-based nongovernmental organizations such as Oxfam, to international financial organizations such as the

World Bank, to local women's movements. As a result, global development campaigns and policy initiatives have increasingly prioritized girls' education, most recently in the Millennium Development Goals (MDGs) and the "Education for All" (EFA) campaign.

One consequence of this interest has been an evolving concern with tracking the progress of "gender equality" in education, a trend illustrated by internationally prominent targets and progress reports. Yet despite a broad consensus over aims, a variety of assumptions concerning gender and development can be detected among the different groups involved. There are a number of ways in which "equality" in education can be defined and demonstrated with educational statistics, with understandings of equality ranging from equal *access* to education, equal *opportunities* within school, equality of *skills* gained, or by other outcomes. As an alternative to human capital evaluations, which chiefly focus on skills and productive capacities, theorists have recently turned to the capability approach as a means of exploring educational equality in terms of individual freedoms.

Measuring Educational Inequalities

The ability to track progress and change in education is important from the point of view of both human capital and human rights. However, assessing education, particularly through the use of numbers, is a controversial issue. In the first place, there are certainly limits to what quantitative data, particularly at macro-level, can reveal about educational processes. In relation to gender in particular, an over-reliance on quantitative data can be at the expense of understanding the differential experiences of girls and women; statistical methods alone are insufficient for exploring complexities relating to gender, and qualitative studies are crucial for understanding individual interpretation and motivation, and dealing with the direct experience of people in specific contexts. Awareness of the need to examine context and individual experience has led to the clarification of qualitative frameworks for measuring inequalities, although the quantitative/qualitative distinction is increasingly being challenged as a false dualism.[1]

A second concern over educational measurement relates to the question of defining equality in education. Quantitative descriptions commonly present "inequality" as an objective reality. Yet in order to understand the limits of existing measurements we must first look more closely at the ways in which inequality can be a social construction, due to differing ideas about which aspects of education require equal distribution (Foster 1996). In particular, defining *gender* equality in schools is far from being an objective task. Several theoretical perspectives exist, underpinned by different assumptions about gender, development, and women's rights. Human capital approaches, for example, have emphasized equal distribution of instrumental skills and the subsequent improvement of individual material welfare (Schultz 1961). In the UK, feminist campaigns to improve girls' schooling in the 1980s became divided over definitions of gender equality in education: whether this should be seen as equality of *access;* equality of *achievement* and *opportunity;* equality of *treatment,* or equality of *outcomes* and gender equity in society (Arnot and Phipps 2003; Weiner 1985). There has been an absence of a single, consensual theoretical framework within which to conceptualize gender equality in education (Erskine and Wilson 1999).

Yet the issue of educational measurements has become particularly pressing in developing countries owing to the growth of particular kinds and forms of evidence-based policy and practice and the assumptions that flow from these within international development policy. Currently, the most visible examples of this are the EFA campaign and the MDGs, both of which contain concrete, time-bound, quantitative targets (World Bank 2003, p. 3).[2] Specific targets at the national level accompany the international goals for primary and secondary schooling: for example, as part of the EFADakar framework (2000), each country was required to produce a national action plan by 2002, incorporating specific, time-bound national education goals.[3] In part, this emphasis on quantitative measurement can be seen as evidence of commitment to the achievement of the MDGs and EFA goals; it also reflects changes in the management theories and governance strategies of international agencies (King 2004; King and Buchert 1999). This trend has provoked concerns that reliance on the existing, simplistic educational indicators may mean that policies have a negligible, or even detrimental effect on educational experiences as reforms become focused on the achievement of narrow goals such as enrollment (Goldstein 2004).

Yet for exploring inequalities, there is a very real need for quantitative data to highlight educational disparities through comparison, and to address them effectively. Gorard notes that almost all education research projects start from some form of quantitative basis (2001, p. 3). Further, the use of numbers enables correlation studies, which can explore links between certain educational characteristics and other contextual variances. On a more pragmatic note, arguments couched in the language of numbers may be more influential: in early feminist educational campaigns in the UK, for example, significant use was made of quantitative data to highlight gender disparities in educational outcomes (Yates 1985; Arnot et al. 1998). To engage fully with current development policy, researchers may have more power if their critiques employ similar methodological tools and if they are expressed in quantitative terms.

International agencies are themselves acutely aware of the limited picture that current measurements reveal about educational experiences.[4] Data availability and quality issues mean that only a handful of indicators, such as school enrollment, survival, or expenditure, are internationally comparable; only a few others, such as dropout rates or numbers of female teachers, make an appearance in international policy documents and reports. Strategies for improving and expanding the range of existing measurements are a common feature in recent campaign documents, but while additional indicators have been suggested, there are more questions concerning the further theoretical exploration of gender equality in education.[5] How should educational equality be conceptualized and evaluated beyond basic access measures? It is with these questions in mind that some educational researchers have recently turned to the capability approach as a framework for understanding and measuring educational inequality.

Conceptualizing Capabilities and Education

Sen's capability approach was developed largely out of concern with the inadequacy of existing methods for measuring inequalities, which tended to be based on either interpersonal assessments that exclusively focused on people's mental states (in terms of happiness or satisfaction), or approaches that focused on physical or financial resources. Instead of focusing on individual happiness and self-evaluated satisfaction or absolute measures of the goods and resources that people access or possess, the capability approach is concerned with "the various things a person may value being or doing," or *functionings*, and the freedom of individuals to achieve these functionings (Sen 1985, pp. 197–198; 1992, pp. 39–53; 1999: p. 75). Functionings, for example, may include working, resting, being literate, being healthy, being part of a community, being respected, and so on. The "freedom" aspect is particularly important because individuals differ over things that are valuable to them. So it should not be assumed that everyone would aim to achieve the same functionings given the same opportunities and freedoms. In real life, therefore, two people with very similar capability sets will probably end up with different types and levels of achieved functionings as they will have made different choices owing to their different desires about the kind of lives they wish to lead (Robeyns 2003, p. 14).

Further, Sen argues that because individuals differ in their ability to convert resources into "doings or beings" that are important to them, providing an equal command over resources does not equate to equal

opportunities. Some of these differences will be due to individual preference, others due to structural differences in society related to gender, class,
race, caste, etc. Sen details an example of the relationship between an individual having access to the commodity of food, and the functioning of "being
well-nourished" (the relationship varies according to a number of factors
such as metabolic rate, body size, age, activity levels, the presence of para-
sitic diseases, etc.) (Sen 1985, pp. 198–199). Similarly, a distinction needs to
be recognized between the presence of a school and the functioning of
"being well-educated." In this way, a similar bundle of commodities will
generate different *capability sets* for different people.

In terms of theories of educational inequality, the capability approach
offers three main advantages over comparisons of levels of access or outcome. First, it provides a wider vision of individual rights to the human
capital focus on economic productive capacities. Second, capabilities, as
opposed to functionings, are able to reflect the importance of individual
autonomy and choice. Last, rather than placing emphasis on resources
available to an individual, the approach takes into account the ability of an
individual to convert resources into functionings.

The capability approach has been employed by researchers to investigate a number of different aspects of educational experiences. The
acknowledgment of individual values as a valid consideration in assessment has provided theoretical space for exploring student aspirations
(Flores-Crespo 2004; Watts 2005). The main body of work so far has
investigated the general relationship between education and expansion
of an individual's overall capability set. Alkire (2002, pp. 255–271) conducted a series of interviews based on a capability framework to assess
the impact of an Oxfam literacy project for women in Pakistan, and used
their responses to outline how a scheme, which would no longer be
funded if evaluated only by traditional economic and financial outcomes,
had other beneficial effects besides an increase in personal income from
greater literacy. Traditional criteria might include the total number of
graduates who have achieved literacy since the start of the project, the
unit cost per graduate, the income generated by the income-generation
component, the projected future socioeconomic benefits of this human
capital improvement, or institutional strengthening. This "cost-effective"
measurement has been previously used to evaluate development projects.
Lorella Terzi has used Sabina Alkire's concept of basic capabilities to
explore which "enabling conditions" might be considered to constitute
education, and to suggest a list of "basic capabilities for educational
functionings" (Alkire 2002; Terzi, chapter 2 of this book). Flores-Crespo
(2004, p. 8) investigated the pedagogical side to capability expansion and
contends that in the sphere of education, capabilities can only genuinely

be enhanced through liberal education because this seeks to "liberate the mind from the bondage of habit and custom" through promoting reason, personal autonomy, and independence. Drawing on Martha Nussbaum's guidelines, Flores-Crespo identifies this as entailing a critical examination of oneself, conceiving of ourselves as world citizens, and a capacity for narrative imagination and empathy. Similarly, Walker has identified a number of "core education capabilities," including practical reason, affiliation, emotional integrity, and knowledge, which can be seen as "capabilities which we might expect education to foster" and without which education would not contribute to "a fairer society, or powerful and positive learning, learner agency and human development" (2004, pp. 8, 16).

Particularly concerned with international measures for comparison, Elaine Unterhalter has used the capability approach to explore the ways in which the experience of attending school may be detrimental to an individual's capabilities. She used the approach to challenge the notion, generally accepted within development frameworks, that schooling is a universally neutral or beneficial part of development, which implies that the evaluation of "functionings and capabilities relating to education can be metanymically accomplished through considering the outputs of schooling, for example levels of literacy or years in basic education" (Unterhalter 2003, p. 8). Formal schooling for girls in South Africa, where there are soaring HIV infection rates and a high incidence of sexual harassment and rape in schools, means that the process of education can reduce the freedoms of women; in these cases, schooling "literally ends the girl's life, destroying her capability" (Unterhalter 2003, p. 16).

How might we theorize about capabilities if we are concerned with the measurement of gender equality in education? Central to the capability approach is the freedom for an individual to achieve functionings, particularly those that are valuable to them. However, a wide range of possible functionings are involved in the process of formal education, for example: attending school, completing class work, learning a new skill, gaining confidence in one's own abilities, passing an exam, developing autonomous thought, gaining employment, or engaging in democratic processes. How does such a broad range of potential actions relate to notions of well-being and agency, and how can they be negotiated for the purposes of comparing and evaluating individual capabilities? Approaches so far have tended to treat the topic of capabilities within education as a complete and homogeneous concept. I would like to suggest here that it might be possible, and even theoretically advantageous in terms of measurement, for education to be conceptualized as occupying two different positions in relation to an individual's capability set.

An important distinction, I believe, is between cases in which the functioning in question is a formal educational process itself, and cases in which functionings have been enabled through formal education. Taking Robeyns's diagram of factors relating to an individual capability (2005, pp. 98–100), it is possible to see that education can occupy different roles depending on which particular capability is being considered. It can be a functioning itself (which I have indicated with "**A**" in figure 6.1), where we are concerned with an individual's capability to achieve this functioning: Does a child have the capability to attend school, or to acquire a new skill, or engage with a new concept? Alternately, formal education can act as a *conversion factor* that determines the extent to which a person can generate capabilities from goods and services (which I have marked with "**B$_1$**" and "**B$_2$**"); for example, the capability to exert a legal right enabled by literacy, or the capability to choose who to vote for by autonomously negotiating political materials and the media.

Formal education can, therefore, be seen as both a functioning in itself that one may or may not have the freedom to achieve (i.e., "being formally educated") and something that enables other possible functionings. In order to conceptualize gender inequalities in education, it might be helpful to think about evaluation in two separate but related spheres: the capabilities of children *within* the educational experience (or the capability

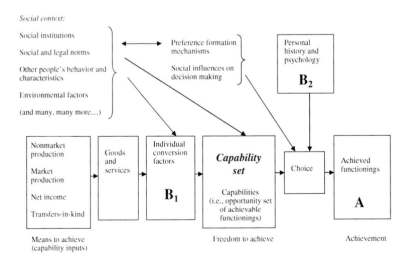

Figure 6.1   The different roles of education in Robeyns's stylized nondynamic representation of a person's capability set and her social and personal context
Source: Based on Robeyns 2005, p. 98.

to participate in education) and capabilities *gained through* education, which are exercised both during and after school years. This distinction is somewhat artificial as I will outline later on and perhaps more complicated and overlapping in practice, but this approach is suggested as one way of negotiating current theoretical difficulties in relation to formal education measurements, and education policy. In particular, I believe that there are important distinctions in the arenas in which well-being and agency can be evaluated.

It is also important at this point to be clear about the difference between education that takes place as part of a process of formal schooling, and a much broader, generalized notion of "education" that takes place outside school in contexts including the household, the economy, the media, and peer groups. It is possible for an individual to learn both instrumental skills and social attitudes in *both* of these arenas. In this chapter, however, taking education policy as my central concern, I address only the notion of formal schooling and the skills and social conditioning within it.

The Capability to Participate in Education

If "being formally educated" is considered as one type of functioning in itself, then the capability to be educated can be defined as the freedom for a child to fully participate in the school-learning process.[6] This, therefore, involves factors that enable a child to attend school; and once physically attending school, being able to participate and understand, and engage in learning confidently and successfully. In a way, this might be considered as the full working of the "mechanics" of the educational process.

An assessment of the capability to participate would, therefore, compare constraints that might affect the freedom of a child to achieve various educational functionings. What might prevent a child from engaging in the learning process, and what enabling conditions are necessary? Factors that might affect the ability to attend school, understand and participate in educational settings are present both outside the school and within the school environment, and they include social institutions, social norms, personal characteristics, and environmental factors. For example, external factors might include the availability of a school, financial issues (if parents can only afford to send one child to school, the child's sibling's capability to participate is curtailed), and household responsibilities (if a child is needed to help at home the child may not be able to enroll, or the child's attendance or full participation may be affected). The capability to participate would also be affected by conditions within the school environment such as personal safety issues, teacher behavior and competence, violence in school, or facilities. (It is possible that this could also include basic skills such as literacy and numeracy, which are non subject-specific and fundamental to learning processes.)

In this sphere we might think of well-being as "(formal) educational well-being," or full participation. If the capability to participate in education is limited, therefore, then the level of educational well-being will exist at a low level. As we are concerned with a functioning relating to a specific area of life—formal education—notions of agency should also be considered in this way. Here, therefore, agency can be understood in terms of the *educational* functionings valued by each individual child, such as the choice of a particular subject, attending a particular school, or attaining a particular level of achievement.

Capabilities through Education

Once children are participating in education, once the "mechanisms" of education are functioning correctly, it is possible to consider the capabilities that can be gained *through* education (or not gained, or even possibly lost). In this case analysis is concerned with the contribution of education to other life functionings, as opposed to purely educational functionings.

Formal education can enable other functionings both directly and indirectly. First, it can facilitate a huge range of specific functionings, such as employment, which requires particular qualifications, understanding health issues, or engaging with civil society and political processes. To measure inequality in this "direct" sphere we might look at the specific skills, subjects, and content of the education that is received. Yet it is problematic to measure exactly the contribution of education to a particular outcome, because many functionings in life are also dependent on other external factors. For example, someone may have the skills for a particular job, but the freedom to perform these skills would depend on the availability of jobs. Second, in addition to its directly instrumental value, we can also think about how formal education can bring greater choice, both through having a broader range of skills available to choose from, and by enabling an individual to reason and think autonomously about the options that are available to them.[7]

To illustrate further the difference between the "capability to participate" and "capabilities through education," we can explore how they might relate to Sen's categories of well-being and agency achievements and freedoms.[8] For the capability to participate in education, well-being is defined as well-being within the formal education process, or as "educational well-being." This involves achievement of functionings considered to be central to educational well-being: attending school, participating in lessons, understanding and learning—overall, full participation in formal education. Similarly, "agency" relates to *educational* functionings that are valued by the individual, such as taking a particular subject, attending a particular school, or achieving to a particular level; it might be considered as "educational agency."

In contrast, for capabilities through education, "well-being" and "agency" can be defined as general well-being and agency functionings in life in general, rather than relating specifically to education. Formal education may contribute directly to the achievement of certain well-being and agency functionings, such as getting a job that requires literacy. Moreover, formal education can contribute to an individual's *freedom* to achieve well-being and agency functionings (or capabilities), both through the range of direct skills that has enabled these freedoms, and through the reasoning and intellectual abilities that are central to freedom and the process of choice.

It is interesting to consider the position of well-being and agency freedoms in a system of free and compulsory education. I would argue that when such a system is fully functioning, these *educational* well-being and agency freedoms only exist at a low level (although it should be noted that they will still have well-being and agency freedoms in other aspects of their lives). In a system of compulsory education, children do not have the freedom to choose whether to attend school or not, or the freedom to choose whether to participate in the learning process or not. If there is a curriculum they may not be able to choose particular subjects, and it is very unlikely that they will be able to choose the political aspects of their education. Probably the main area they have freedom in is in the amount of effort that they devote to learning. Sen himself has stated that while it may be the case that freedoms are constricted under compulsory education, the future freedoms of the child should be considered (Saito 2003). The picture would, however, be considerably different for higher or adult education.[9]

The distinctions between well-being and agency in the two categories are further illustrated in table 6.1. The first column represents the capability to participate in education and the second, capabilities gained through education.

Clearly, there are limits to how far the distinction between the capability to participate and capabilities through education can be taken. If a child is not able to participate fully in education, then this will affect the capabilities that can be gained through education. But it is important to recognize that education occupies a different place in an individual's capability set according to the inequality concerned—it can either be the potential functioning that is being inhibited, or it can be a conversion factor that enables other capabilities and that substantially different aspects of well-being and agency are the focal point in each case.

Table 6.1 Well-being and agency achievements and freedoms within, and through, formal education

	Capability to participate in education	*Capabilities gained through education*
Well-being achievement	("Educational well-being") Attending, participating, understanding. Can include basic educational skills such as literacy and numeracy	Contribution of education to other capabilities crucial to general well-being: e.g., employment, health
Agency achievement	("Educational agency") Educational aspects that are valued by the individual (e.g., attending school, completing schoolwork, level of achievement in participation, attending a particular school, studying a particular subject)	Contribution of education to valued functionings and capabilities: e.g., employment sector, valued level of health, political engagement, family life
Well-being freedom	Ability to choose to attend, participate, understand Barriers and constraints to this freedom in compulsory education	Contribution of education to the freedom to achieve well-being and valued functionings 1) Having a range of skills and, therefore, options
Agency freedom	Ability to choose to achieve aspects of education valued by the individual Barriers and constraints to this freedom in compulsory education	2) Reasoning and autonomous thought; choice and preference formation: —finding out which options are available —ability to reason about options —knowledge of how to overcome constraints to options

The last section of this chapter turns to the question of how such a framework might be applied if the specific concern is with gender equalities in and through education. The focus of this chapter is on the problems that girls may encounter in relation to formal education, largely because measurements relating to female education have become a concern in recent development campaigns; but it would be equally illuminating to consider the effect of schooling on boys' capabilities.

Gender Equality: Particular Concerns

This chapter has so far argued that when evaluating freedoms in relation to education, it is necessary to make a theoretical distinction between constraints to the capability to participate in formal education, and constraints that children face due to the nature of the formal education they receive or, conversely, how education can remove constraints that they might otherwise face as adults. If we are concerned specifically with gender inequalities in capabilities, there are important differences in how inequalities should be evaluated in these two spheres. (Of course we might apply the argument to other structures of inequality, such as race, or social class, or language as well, but here I turn illustratively to gender inequality).

First, considering whether there is gender equality in the capability *to participate* in education, it would be necessary to scrutinize if there are similar barriers and constraints or enabling factors to educational well-being and educational agency for boys and girls. Given the low level of agency and well-being freedoms in a system of compulsory education, it might be possible to compare levels of achieved functionings, such as boys' and girls' attendance, completion, and achievement levels.

If we are specifically concerned with gender equality, how might we approach the second category, capabilities gained through formal education? Education is widely regarded as a beneficial force in development at both the individual and macro level, and in the last 20 years girls' education in particular has been highlighted as important for a number of reasons. Not only has female schooling been consistently linked to a number of positive social and economic effects, but it is also seen as central to women's rights in view of its perceived ability to redress structural imbalances by providing women with the means to work toward their own goals and overcome the circumstances they were born into (McMahon 1999; Schultz 1961; World Bank 2001). Liberal educational theorists have typically taken the view that allowing girls to enter the formal school system can only work toward bringing about both equality of educational experiences, and of gender in wider society (for further discussion see Dillabough and Arnot 2002). In a significant amount of literature, including capabilities-influenced human development approaches, the "human capital" and "rights" approaches to female education are seen as entirely compatible, with simultaneous benefits at macro- and microeconomic level, and in terms of individual rights. Sen himself outlines that increased personal income and economic growth can expand opportunities and are, therefore, valuable outcomes of education, although they have to be "integrated into that more foundational understanding of the process of development as the expansion of human capability to lead more worthwhile

and more free lives" (Sen 1999, pp. 292–297). A number of human development theorists have argued that regardless of content, basic tools
imparted through education such as literacy and numeracy will have an
overall beneficial effect on the individual. Drèze and Sen state, "The patriarchal orientation of school textbooks . . . does not detract from the fact
that an educated woman is better placed to liberate herself from the economic dependence on men that shackles so many Indian women, to gain
independent access to information, and to make her voice heard within the
family" (Dreze and Sen 1995, p. 145; PROBE 1999).

However, critical theorists argue that education can actually exacerbate
persisting inequalities (Bowles and Gintis 1976; Shukla and Kaul 1998;
Willis 1977). From this perspective, formal education has the potential to
reproduce inequalities, both through differential provision (the availability
of quality schooling may be an option only available to those with a certain
level of income), and through content and pedagogy (teacher and peer
behavior may encourage certain trajectories). In relation to girls' schooling,
radical feminist educationalists have argued that attending school does not
inevitably challenge gender inequality, and can actually serve to perpetuate
imbalances, particularly through pedagogy, content, and teacher expectations (Arnot 2002; Weiner 1994).

In developing countries, feminist authors have emphasized these claims
through critical assessments of development policies on girls' education,
highlighting in particular a preoccupation with fertility and the reproductive and productive aspects of women's lives (Jeffery and Basu 1996;
Stromquist 1995). From a feminist viewpoint, assessments of gender equality in schools should take into account the levels of individual freedom
and empowerment that girls gain from their education, reflected in both
achievements and outcomes from education. Kabeer (2003, pp. 131–133)
outlines concerns with situations where education is seen as equipping girls
to be better wives and mothers, or increasing their chances of getting a suitable husband. In these situations, education does little to equip girls and
women to question the world around them and the subordinate status
assigned to them. Education may thus increase women's effectiveness in
their traditional roles but is unlikely to override, and may even reinforce,
restrictive interpretations of these roles. The *content* of education can often
mirror and legitimate wider social inequalities through gender stereotyping
in the curriculum, restrictions in subject choices, and teacher attitudes
(Jayaweera 1997; Longwe 1998; Subrahmanian 2002).

In the light of these debates, how might capabilities gained *through*
formal education be conceptualized? First, we might think about particular
instrumental *skills* (e.g., for specific functionings, such as a particular job).
A broad definition of skills would encompass basic skills such as literacy,

numeracy, and social and communication skills, as well as more specific, subject-based ones; in many ways this is closely linked to a human capital approach. This particular aspect mainly relates to B_1 in figure 6.1, as a factor that allows the conversion of goods and services to a capability. One example might be converting a job vacancy to a job that a woman is capable of doing; or at a more basic level, converting a newspaper into something through which a woman can understand and learn about the day's events. Equal capabilities through education depend, therefore, on whether there is equality in receiving these elements (e.g. whether equal numbers of boys and girls take a course in physics and if equal numbers pass the physics examinations with a similar spread of achievement levels). However, as mentioned earlier, levels of gender equality in skills received does not reflect whether an individual has the freedom to exercise these skills or not—for example, whether equal numbers of qualified men and women will be offered a job as a particle physicist—owing to wider social and economic factors.

Second, particularly in relation to freedoms and capabilities through education, we might consider how education relates to *choice* and *preference formation*. This is a particularly important consideration from a gender perspective. An individual will have a greater number of options if they have a wider range of skills. But education can also affect how individuals reason about the options that might be available to them and affect the choices they make, particularly in relation to following any valued goals they may have for the future. The ability of education to act as an enabling factor through expanding choice in this way is demonstrated in figure 6.1 by B_2.

How should this be theorized in relation to gender inequalities in wider society? An examination of social constraints and influences on choice is central to a *feminist capability perspective* (Robeyns 2006). Can girls be held fully responsible for the achieved functionings they choose? A feminist capability perspective includes two main points: first, that for many valuable capabilities, the constraints for women are larger. So, for example, a woman may have chosen to work as a part-time nurse, but it is possible that gender discrimination, segregation on the labor market, and gendered distribution of household tasks have influenced this choice. Second, in addition to gendered constraints on choice, actual preferences underlying the choice are shaped by gender socialization and other gendered social and psychological processes. So a woman's choice to become a nurse may have been influenced by socially prevalent notions of femininity and the suitability of "caring" professions for women.

Therefore, a further aspect of education needs to be considered in addition to skills imparted. Schooling may be gender-neutral and not emphasize male/female stereotypes; but students will nevertheless encounter expectations and pressure to conform to particular gender roles and identities in wider society both during and after their years of formal schooling. Certain curricula, however, have the potential to make students actively aware of the gendered nature of society, and the nature of the constraints they may face, and moreover equip them with strategies and skills to follow their own goals regardless of gender-role expectations. An education process that explicitly imparts a conception of gender relations based on equality can empower female pupils—both through giving them skills and competencies that would normally be denied to them and through a self-belief in challenging gender constraints that they will face on entering the world as adults, and through knowledge about the structural constraints needed to challenge them. In the same way, education can confirm existing inequalities in gender relations through imparting messages encouraging the adoption of restrictive gender role-models, such as typically "female" career paths or life choices. It could be argued that educating girls in this way does not expand their capabilities in the same way as an education that has a transformatory gender content, which aims to reduce the constraints girls will face in society by expanding their ability to follow goals that are free from gendered expectations.

Therefore, an equally important aspect comes through the content of education, in terms of any gender-relevant aspects of the education received. I have termed these *gender-related educational capabilities* and they pertain to the gender values imparted through the education process. These would, therefore, include whether gender stereotypes exist in teaching and curriculum materials; whether education includes learning about the gendered nature of society; whether it provides preparation (skills and knowledge) to take on gender-based problems in life after school, and therefore working against adaptive preference; and the ways in which a school can act as an exemplary institution in terms of gender, such as gender relations in school, staff ratios, differential treatment of boys and girls, safety.

It is possible to compare these gender considerations with the general considerations for equality of capabilities gained through education. For well-being achievement, this would mean boys and girls gain equal capabilities from education toward achieving well-being functionings such as health, employment; for agency achievement, boys and girls would gain equal capabilities from education toward achieving valued functionings such as type of job or political involvement. Similarly, for well-being and agency freedoms, it would mean no gender inequality in the contribution of education to the freedom to achieve well-being and valued functionings.

Table 6.2 outlines the distinction between the general considerations for equality through education, and considerations that relate to gender equality in particular. The first column shows factors to be considered for

Table 6.2   General capabilities through education, and gender considerations

	Capabilities gained through education	Capabilities gained through education: gender perspective
Well-being achievement	Contribution of education to other capabilities crucial to general well-being: e.g., employment, health	Gender-equal contribution of education to other capabilities crucial to general well-being: e.g., employment, health
Agency achievement	Contribution of education to valued functionings and capabilities: e.g., employment sector, valued level of health, political engagement, family life	Gender-equal contribution of education to valued functionings and capabilities: e.g., employment sector, valued level of health, political engagement, family life
Well-being freedom	Contribution of education to the freedom to achieve well-being and valued functionings	Contribution of education to the freedom to achieve well-being and valued functionings free of gender constraints
Agency freedom	1) Having a range of skills and, therefore, options 2) Reasoning and autonomous thought; choice and preference formation: — finding out which options are available — ability to reason about options — knowledge of how to overcome constraints to options	1) Having an equal range of skills between males and females 2) Reasoning and autonomous thought; choice and preference formation. How does education affect choice and preference formation along the lines of gender? = "gender related educational capabilities"

general capabilities gained through education, reproduced from table 6.1. The second column outlines specific considerations if the concern is with gender equality.

It is interesting at this point to consider the relationship between education and values. While the capability approach recognizes individual values and preferences as a central part of an individual capability set, education can undeniably have a role in *imparting* values (such as perceptions about what gender roles should specifically constitute) and preferences. This point is particularly pertinent in relation to gender equality and education: if education can encourage an individual to desire a particular role or position in society, to what extent can the idea of gender equality be seen as a *value* itself, as opposed to a normative framework of justice?

This rests on the understanding of gender equality in society, which determines notions of injustice and inequality: different interpretations of gender inequality in schools reveal implicit assumptions about the nature of difference between girls and boys. For example, a focus on disparities in achievements is only possible if it is understood that girls and boys overall have equally balanced academic abilities. In the same way, a focus on formal outcomes such as labor market participation is informed by the belief that women and men have both equal abilities and aspirations in the workplace.

The concept of equality of capabilities is central to the capability approach; however, Robeyns (2006) has observed that the capability approach does not contain a specification of gender equality and can, therefore, accommodate a more conservative definition of gender equality that rests on the belief that men and women have inherently different preferences. This chapter is situated in the belief that gender is a social construction and that girls and boys should have equal and identical rights within and through education; a belief in essential differences between boys and girls can lead to very different views about gender and education (Tooley 2002). In the absence of an additional normative theory of gender relations, therefore, a capability assessment of schooling might not regard differing outcomes for boys and girls as a matter of injustice.

A full discussion of how this framework might be made operational in practice is not possible here. But it is worth considering the implications of this framework (the division between capability to participate and capabilities through education) for actual measurement and comparison. In practice, many problems surround the empirical measurement of educational capabilities. First, problems with current measures for the capability to participate in education, particularly at the macro level, limit what can be said about capabilities through education and the gendered content of education. However, I believe it is not pointless to look at the gendered nature of educational content even without full participation measurement; it is unlikely to have *no* effect unless girls are not attending school at all. Second, in terms of collecting evidence for gender-specific education capabilities, classroom studies are needed to look at the gendered content and nature of education. Third, as with all capability assessments, it is not possible to achieve a broader gender-education capability assessment without individual studies relating to values or to options available; and this is only possible at the micro level; macro-level outcomes are currently limited to measuring only achieved functionings.

These are all considerations that would relate to an attempt to create additional capability-based measures for education. How does this framework affect our understanding of *existing* educational data such as enrollments, survival rates, and examination results? I would argue that the framework outlined in this chapter suggests how current variables might be interpreted and prioritized in a capability-centered evaluation.

Enrollment rates, for example, can be seen as an indicator for the capability to participate: that an initial level of educational well-being is being achieved; survival rates are an important additional measure. Measuring agency achievement in education participation, however, would involve some level of information on valued functionings at the individual level. Similarly, for capabilities through education, gender parity in examination results may suggest greater equality in the distribution of skills (individual conversion factors, B_1), but this constitutes only one aspect relating to an individual's overall well-being and agency.

Conclusion

This chapter has attempted to provide a theoretical framework for gender equality in education in terms of individual capabilities. I have argued that it is possible to be more specific about how measurements relate to capabilities if we are clear about whether the concern is with education as a functioning in itself (in which case capabilities to achieve this functioning must be compared), or as a conversion factor toward other capabilities (when the content and processes of formal schooling need to be compared for their enabling effects). In particular, restrictive gender role reproduction through education can be viewed as a constraining factor on more general capabilities in other areas of individual functioning.

The question of measuring progress in education remains an important one in the context of increased policy focus on girls' education and the use of targets within international development policy. Recently, some theorists have questioned whether the Millennium Development Goal of increasing girls' enrolments always leads to an increase in their capabilities, and while it is not my intention to argue against increasing the numbers of girls in school, important questions remain concerning the relationship between rising female enrolment rates and the freedoms experienced by school-age girls. The capability approach offers valuable insights as a framework through which these issues can be explored.

Although the chapter was written in the context of the prevailing development preoccupation with girls' education in developing countries, the theoretical framework extends beyond this remit. In imparting messages about gender in general, formal schooling can equally have an impact on boys' capabilities through the reproduction of masculine stereotypes, or the ability to challenge them. Further, while levels of inequality in the capability to participate in formal education are currently starkest in the developing world, the concept of capabilities through education can be used as a lens for scrutinizing inequalities in the education systems of both developed and developing countries.

Notes

1. See Leach 2003. For discussion about the distinction between qualitative and quantitative methods in educational research, see Gorard (2001, p. 6).
2. Although some earlier campaigns included specific targets, see table 5.4 (UNESCO 2003, p. 197).
3. For example, Pakistan's National Action Plan sets targets for net female enrollment in primary education: 6.5 million in 2005, 7.57 million in 2010, and 8.5 million in 2015. (British Council, Pakistan Ministry of Education, and NORAD 1993, p. 107).
4. For example, the 2002 *Report of the Inter-agency and Expert Group on MDG Indicators* acknowledges that "enrolment does not imply achievement of primary education, which is why it is complemented by indicators of survival and literacy. But even with these additional measures, it provides only limited information on achievement" (UN Secretariat Statistics Division 2002: 6).
5. For one example of recent discussions over indicator improvements, see UN Millennium Project 2004.
6. "Being informally educated," or being educated through other life experiences are other potential functions related to learning; however, as mentioned above, the focus of this chapter is education processes within schools.
7. This distinction between the capability to participate in education and capabilities gained through education can be related to Des Gasper's (2002) identification of two definitions of capabilities: *S-caps* or "skills capabilities," and *O-caps* or "option capabilities," which depend on environmental factors. When thinking about the capability to participate in education, we might chiefly be concerned with O-caps, or external factors; whereas capabilities gained through education can be linked to the notion of S-caps. Gasper's O- and S-caps categories, however, are offered to clarify what he sees as an "underelaboration" of the general concept of freedom in Sen's work; here, my concern is merely to illustrate how it is necessary to be specific about the different roles of education when concerned with freedoms.
8. Sen has located four conceptual spaces within which human life can be evaluated: well-being achievement, well-being freedom, agency achievement, and agency freedom. Alkire summarizes these: "Well-being achievement refers to the achievement of those things that are constitutive of one's well-being; and well-being freedom is one's freedom to achieve these things. Agency achievement refers to a person's success in the pursuit of the totality of their considered goals and objectives; and agency freedom is one's freedom to bring about the achievements one values and which one attempts to produce" (Alkire 2002, pp. 129–130).

9. Adult and higher education, being noncompulsory and with the curriculum generally chosen by the student, potentially involve a much higher level of well-being and agency freedoms.

References

Alkire, Sabina. 2002. *Valuing freedoms: Sen's capability approach and poverty reduction* Oxford: Oxford University Press.
Arnot, Madeleine. 2002. *Reproducing gender? Essays on educational theory and feminist politics.* London: Routledge.
Arnot, Madeleine, John Gray, Mary James, and Jean Rudduck. 1998. *Recent research on gender and educational performance.* London: OFSTED.
Arnot, Madeleine, and Alison Phipps. 2003. Gender and education in the UK. In UNESCO EFA Monitoring Report 2003–04.
Bowles, S., and H. Gintis. 1976. *Schooling in Capitalist America.* New York: Basic Books.
British Council, Pakistan Ministry of Education, and NORAD. 1993. Workshop on female access to primary schooling in Pakistan, February 20–22, 1993, Islamabad.
Dillabough, Jo-Anne, and Madeleine Arnot. 2002. Feminist perspectives in sociology of education: Continuity and transformation in the field. In *Education and sociology: An encyclopedia,* edited by D. Levinson, A. Sadovnik, and P. Cookson. London: RoutledgeFalmer.
Dreze, Jean, and Amartya Sen. 1995. *India: Economic development and social opportunity.* Oxford: Oxford University Press.
Erskine, Sheila, and Maggie Wilson, eds. 1999. *Gender issues in international education: Beyond policy and practice.* London: Falmer Press.
Flores-Crespo, Pedro. 2004. Situating education in the human capabilities approach. Paper read at the Fourth International Conference on the Capability Approach, September 5–7, at Pavia, Italy.
Foster, V. 1996. Space invaders: Desire and threat in the schooling of girls. *Discourse: Studies in the Cultural Politics of Education* 17:43–63.
Gasper, Des. 2002. Is Sen's capability approach an adequate basis for considering human development? *Review of Political Economy* 14 (4): 435–461.
Goldstein, H. 2004. Education for All: The globalization of learning targets. *Comparative Education* 40 (1): 7–14.
Gorard, Stephen. 2001. *Quantitative methods in educational research: The role of numbers made easy.* London: Continuum.
Jayaweera, Swarna. 1997. Women, education and empowerment in Asia. *Gender and Education* 9 (4): 411–423.
Jeffery, Roger, and Alaka Basu, eds. 1996. *Girls' schooling, women's autonomy, and fertility change in South Asia.* New Delhi: Sage.
Kabeer, Naila. 2003. *Gender mainstreaming in poverty eradication and the millennium development goals.* Canada: International Development Research Council.
King, Kenneth. 2004. Development knowledge and the global policy agenda. Whose knowledge? Whose policy? Occasional paper, Centre for African Studies,

University of Copenhagen.http://www.teol.ku.dk/cas/nyhomepage/mapper/
Occasional%20Papers/King_%20Samlet.pdf.
King, Kenneth, and Lene Buchert, eds. 1999. *Changing international aid to education: Global patterns and national contexts.* Paris: UNESCO.
Leach, Fiona. 2003. *Practising gender analysis in education.* Oxford: Oxfam.
Longwe, Sara. 1998. Education for women's empowerment or schooling for women's subordination? *Gender and Development* 6 (2): 19–26.
McMahon, Walter. 1999. *Education and development: Measuring the social benefits.* Oxford: Oxford University Press.
PROBE. 1999. *Public Report on Basic Education in India.* New Delhi: Oxford University Press.
Robeyns, Ingrid. 2003. The capability approach: An interdisciplinary introduction. Paper read at the Third International Conference on the Capability Approach, September 6–8, at Pavia, Italy. Available from www.ingridrobeyns.nl.
———. 2005. The capability approach: A theoretical survey. *Journal of Human Development* 6 (1): 93–114.
———. 2006. Three models of education: Rights, capabilities and human capital. *Theory and Research in Education* 4 (1): 69–84.
Saito, Madoka. 2003. Amartya Sen's capability approach to education: A critical exploration. *Journal of Philosophy of Education* 37 (1): 17–33.
Schultz, Theodore. 1961. Investment in human capital. *American Economic Review* 51:1–17.
Sen, Amartya. 1985. *Commodities and capabilities.* Amsterdam: New Holland.
———. 1992. *Inequality re-examined.* Oxford: Oxford University Press.
———. 1999. *Development as freedom.* Oxford: Oxford University Press.
Shukla, Sureshchandra, and Rekha Kaul, eds. 1998. *Education, development and underdevelopment.* New Delhi: Sage.
Stromquist, Nelly. 1995. Romancing the state: Gender and power in education. *Comparative Education Review* 39 (4): 423–454.
Subrahmanian, Ramya. 2002. Gender and education: A review of issues for social policy. In *UNRISD Social Policy and Development Programme,* Paper No. 9. http://www.unrisd.org/unrisd/website/document.nsf/0/0A8ADED14E7E1595C1256C08004792C4?OpenDocument.
Tooley, James. 2002. *The miseducation of women.* London and New York: Continuum.
UN Millennium Project. 2004. *Task force 3 interim report on gender equality.* www.millenniumproject.org (accessed in June 2004).
UN Secretariat Statistics Division. 2002. *United Nations millennium development goals data and trends 2002: Report of the inter-agency and expert group on MDG indicators.* New York: Statistics Division, UN Secretariat.
UNESCO. 2003. *Gender and Education for All: The leap to equality.* Paris: UNESCO.
Unterhalter, Elaine. 2003. The capabilities approach and gendered education: An examination of South African complexities. *Theory and Research in Education* 1 (1): 7–22.
Walker, Melanie. 2004. Insights from and for education: The capability approach and South African girls' lives and learning. Paper read at the Fourth

International Conference on the Capability Approach, September 5–7, Pavia, Italy.

Watts, Michael. 2005. What *is* wrong with widening participation in higher education. Paper read at Capabilities and Education Network, May, at Cambridge.

Weiner, Gaby. 1994. *Feminisms in education.* Buckingham: Open University Press.

———. ed. 1985. *Just a bunch of girls: Feminist approaches to schooling.* Milton Keynes: Open University Press.

Willis, Paul. 1977. *Learning to labour.* Farnborough: Saxon House.

World Bank. 2001. *Bangladesh-Female Secondary School Assistance Project-II.* Washington: World Bank.

———. 2003. *Gender equality and the millennium development goals.* Washington: World Bank.

Yates, L. 1985. Is girl-friendly schooling really what girls need? In *Girl Friendly Schooling,* edited by J. Whyte, R. Deem, L. Kant, and M. Cruickshank. London: Methuen.

Part II

Applications of the Capability Approach in Education

Editors' Introduction

In this part the authors take up ideas from the capability approach and apply them in educational contexts and settings. It is perhaps not surprising that three of the chapters in this section have gender equity concerns as a focus, given the pervasive problem of girls' and women's exclusion from and marginalization in education internationally. Even in developed countries such as the United States, particular groups of women find themselves marginalized in relation to education opportunities. Nonetheless, while the focus here is mostly on gender equity, we would argue that the capability approach is equally robust in taking up other forms of difference such as race and ethnicity, disability, social class, age, and so on. What is important about the chapters in this section is the international range and diversity of education sectors that are addressed, such that we might argue that the capability approach has resonances across all formal and informal education settings and internationally. It is helpful then to read the chapters that follow as illustrative examples of the way in which to apply the capability approach to education. They by no means cover the education ground exhaustively, but they do point to potential applications and ways forward. They further demonstrate the power of empirical voices brought to bear on theorizing, and the importance of practical applications of the capability approach in education.

Taken together, the chapters in Part II draw on the capability approach as a framework and criterion for equality and social justice in education. Education within a human development and capability approach paradigm is to be evaluated according to how it expands individual freedoms to accomplish what people have reason to value. The specific role of education is then to increase students' freedom in the directions they reflectively value for their well-being and agency. While we cannot guarantee the life choices students will make beyond formal education, education ought to make it at least possible for students to act on the future differently and to renew the common world. As Hannah Arendt (1977, p. 192) writes, "The problem is simply to educate in such a way that a setting-right [of the world] remains actually possible, even though of course, it can never be assured." We ought to provide the conditions—"educate in such a way"— that educational development that supports human flourishing is enabled.

Following the capability approach would require equality of educational capability for diverse students across the world—for girls, for working class women, for poor and disadvantaged children—and not just for those whose family and socioeconomic backgrounds and cultural capital are taken for granted in education. It would further involve the social and educational conditions that support capability development, including, as Richard Bates points out, educational leadership and a democratic learning society. What matters is how education as capability is distributed, and to whom, and whether valued capabilities are developed fairly in and through education. In Mario Biggeri's study we see evidence of poor children indicating what capabilities they take as valuable in their lives, and one of these is education. We need then to ask whether they are able to access the education they desire, and further, how formal and informal education are enabling them to develop their capabilities. Or is that some children and students get more opportunities to convert their resources into capabilities than others? Importantly, the capability development of the learners in these chapters shapes capability, reflexivity, and agency for their future choices and aspirations. Capability now has lifelong effects for each learner.

The first chapter in this section by Richard Bates usefully rehearses again the key features of the capability approach and brings the ideas to bear on the specific issue of educational administration and leadership, locating the latter in a contemporary climate of the erosion of trust in public life. It is important that institutions as much as individuals bear the responsibility for the development of capabilities; the administration of such institutions is then at issue. Leadership for capability development must therefore be grounded in a capabilities curriculum and an appropriate pedagogy rooted in social justice concerns.

The following two chapters both take up capability lists in order to scrutinize gender equality in girls' lives and choices. Such lists as noted in chapter 1 are contested. Notwithstanding such debates and the differing positions on this taken by Amartya Sen and Martha Nussbaum, Melanie Walker demonstrates in her chapter a kind of grounded approach to developing a gender equity list for South African girls. Drawing on ideas from South African education policy, from other capability lists, and from the voices of girls themselves, she generates a draft list for further debate and dialogue. Janet Raynor, in her chapter, draws on Nussbaum's list of central human functional capabilities in relation to what adolescent girls in Bangladesh say being educated means to them. She argues that Nussbaum's list is potentially very useful for identifying the role of education in fostering girls' well-being and agency, or for identifying the reasons for the failure of education to nurture the capabilities of girls. Both these chapters suggest that for practical purposes it may well be important to identify core education capabilities that we might reasonably consider to be important prerequisites for girls' flourishing. Of course, how we do this without being paternalistic and reducing people's agency, or pronouncing from lofty heights on desirable capabilities, is tricky and awkward but this is no reason not to attempt it.

Mario Biggeri's chapter turns to the voices of children internationally and their perspectives on education and what they value. Like the two previous chapters, Biggeri argues that putting the capability approach into practice does have to grapple with the issue of selecting a list of valuable capabilities, albeit not one which is universal. We are reminded here of Severine Deneulin's argument that the capability approach "has to take a position about what are the fundamental features which make a human life worth living if it is to provide theoretical insights for undertaking actions towards well-being enhancement" (Deneulin 2006, p. 31). It follows, she argues, that the freedom to choose valued capabilities is not the only intrinsic good but must acknowledge other factors as well, key to which must surely be an understanding that *education* processes and outcomes ought to enhance freedom, agency, and well-being and not diminish them.

The final chapter in this section by Luisa Deprez and Sandra Butler takes up the access and participation of mature women learners from low socioeconomic backgrounds to higher education, and the significant impact this has on their ability to enhance their own freedoms and widen the genuine choices available to them. The women in this study are all mothers and hence their capabilities will have an impact beyond their own individual lives. The chapter demonstrates that the capability approach can be taken up for education in developed countries and for the postsecondary sector. Deprez and Butler show how education is a major source of these women's empowerment, self-esteem, and well-being; it opens up economic opportunities and new friendships, and enriches their lives and their engagement and participation in their communities. In other words, access to and participation in higher education substantially enhances these women's freedoms.

References

Arendt, Hannah. 1977. *Between past and future*. Harmondsworth: Penguin Books.
Deneulin, Severine. 2006. *The capability approach and the praxis of development*. Houndmills, UK: Palgrave Macmillan.

7

Developing Capabilities and the Management of Trust

Richard Bates

This chapter uses the lens of what is described as the field of educational administration, or school management and leadership, to consider what the capability approach offers theoretically to our understanding of social justice and ethical practice in education. Following Sen, such social justice is seen to be a matter of arranging our social commitments in ways that enhance our individual freedom to live a valued life. Capabilities, and the enhancement of individual capabilities through social arrangements, are then at the heart of the issue of social justice, including values and practices of educational leadership, public life, and institutions. The argument developed in the chapter, in its concern with the development of a democratic learning society founded on the principles of liberty, equality, justice, and democracy, is relevant well beyond the specific issue of educational administration and leadership.

Educational administration has a disreputable history. Originating in the early twentieth century from the coalescence of the municipal reform movement, the cult of professionalism, and scientific management (Bates 1983) it has, from the beginning, been concerned with efficiency, accountability, and control. Despite the attempts of John Dewey, William H. Kilpatrick, and George S. Counts to shape a commitment to education that rested on the values of community, democracy, and social progress, educational administration continued its pursuit of a value-free science of management throughout the century. Despite damning criticisms such as Raymond Callahan's *Education and the Cult of Efficiency* (1962) and Arthur Wise's *Legislated Learning: The Bureaucratization of the American Classroom* (1979), the pursuit of a science of educational administration continued unhindered by any educational, social, or ethical concerns. Despite the attempts of Greenfield (1975), Foster (1986), and others (Bates 1980; 1983) to redefine educational administration as a normative and cultural process concerned with the management of knowledge, culture, and life chances, mainstream educational administration (Boyan 1988) ignored both criticisms and opportunities. At the end of the century, as at the beginning, administration in education, as elsewhere, was concerned with performativity (Ball 1994).

Olssen, Codd, and O'Neill (2004) catch the tenor of the times in their account of how schools became increasingly preoccupied with recording and reporting on life in schools so that:

> Efficiency not only had to be done, but it had to be seen to be done. Efficiency was to be continually demonstrated through the incessant production of records and reports. Educational cost accounting became the order of the day. Teachers were required to keep records, accounting for every hour and every day of the week. Administrators were forever occupied in writing reports and policy statements. Needless to say, there was less and less time for teaching, and schools became places of tedium, ritualistic order and bland routine. Ironically, they became less and less efficient in an educational sense . . . The cult of efficiency had become the cult of managerialism.
>
> (p. 191)

In its contemporary form such managerialism has replaced the mantra of "efficiency" with that of "quality":

> Quality has become a powerful new metaphor for new forms of managerial control. Thus, in the pursuit of quality, educational institutions must engage in "objective setting", "planning", "reviewing", "internal monitoring" and "external reporting". Policy formation and operational activities must be clearly separated. Governance, management and operations are all distinct functions assigned to different roles. The quality of education is reduced to key performance indicators, each of which can be measured and reported.
>
> (Olssen, Codd, and O'Neill 2004, p. 191)

If, at the beginning of the twentieth century, the separation of conception from execution was a pillar of Taylor's (1911) scientific management designed to extract the maximum amount of production from even the stupidest worker, at the end of the century neoliberal management was designed to shackle the normative and potentially subversive interests and enthusiasms of service workers (including those who used to be called professionals) in order to guarantee the standardized performance of allocated tasks:

> Neoliberal policy strategies are founded upon a conception of the person that is self-serving, competitive, and likely to be dishonest. It is a conception

that underpins proposals to separate policy formation and advice from policy implementation, or the separation of funder from provider. In this, while neoliberalism values efficiency, effectiveness and control, it devalues interpersonal trust.

(Olssen, Codd, and O'Neill 2004, p. 92)

Indeed, the devaluing of personal and institutional trust is seen by many to be characteristic of contemporary society. As O'Neill (2002) suggests, it is now trusting that seems risky and hard so that we repeatedly read about untrustworthy politicians and officials, or hospitals and exam boards, or companies and schools. She writes that we "face a deepening crisis of trust" (2002, p. 4) in which the response is more accountability to stakeholders. But, she asks, "can a revolution in accountability remedy our 'crisis of trust?'" (2002, p. 4). Her answer is a pretty clear "No." The answer is "No" because of what seems to be a built-in paradox. Accountability procedures are designed to increase not only efficiency but also reliability, not only to eliminate unnecessary expense but also guarantee performance:

> The diagnosis of a crisis of trust may be obscure: we are not sure whether there is a crisis of trust. But we are all agreed about the remedy. It lies in prevention and sanctions. Government, institutions and professionals should be made more accountable. And in the last two decades, the quest for greater accountability has penetrated all our lives, like great draughts of Heineken, reaching parts that supposedly less developed forms of accountability did not reach.
>
> (O'Neill 2002, p. 45)

Inevitably, such accountability involves greater regulation and greater intrusion into the activities of institutions, agencies, and individuals. She argues:

> For those of us in the public sector the new accountability takes the form of detailed control. An unending stream of new legislation and regulation, memoranda and instructions, guidance and advice floods into public sector institutions. . . . Central planning may have failed in the Soviet Union but it is alive and well in Britain today. *The new accountability culture aims at ever more perfect administrative control of institutional and professional life.*
>
> (O'Neill 2002, p. 46, emphasis added)

Ostensibly this ever more perfect control is designed to make institutions, firms, and professionals more transparent and directly accountable for their actions: to provide guarantees of service to the public. One problem, however, is that the multiplication of agencies and accountabilities produces precisely the distractions and confusions that Callahan and Wise described. Indeed, as Wise (1979) suggests, such multiplication of what may be individually defensible but are, in practice, often mutually contradictory accountabilities produces a "hyper-rationalization" that confuses and overburdens institutions and individuals alike. We are repeating the mistakes that history should have taught us to avoid.

But this is only one of the paradoxes. The other is that while such accountability is presented as increasing accountability of institutions and individuals *to the public*, in reality it does no such thing:

> In *theory* the new culture of accountability and audit makes professionals and institutions more accountable *to the public*. This is supposedly done by publishing targets and levels of attainment in league tables, and by establishing complaint procedures by which members of the public can seek redress for any professional or institutional failures. But underlying this ostensible aim of accountability *to the public* the real requirements are for accountability *to regulators, to departments of government, to funders, to legal standards*.
>
> (O'Neill 2002, pp. 52–53, emphasis added)

Regulators, departments of government, funders, and legislative authorities are themselves institutions in which the public has limited trust. Who guards these guardians is not a question with an obvious answer.

Publics and Leadership

Even if there was some straightforward way to make institutions accountable to the public, another problem presents itself. There is not in fact one public but many. The notion of the "public sphere" is subject to quite considerable theoretical and practical (political) controversy around this issue. As McKee (2004) points out, arguments between modernists (who advocate a single public sphere so as to ensure social and cultural integrity) and postmodernists (who advocate multiple interacting public spheres so as to acknowledge social and cultural diversity) are alive and well. Here the question of "voice" seems to be important. Who has the right to be heard? And in this question we arrive at the heart of the issue of democracy, accountability, and social justice:

> The issue of "democratic inclusiveness" is not just a quantitative matter of the scale of the public sphere or the proportion of the members of a political community who may speak within it. While it is clearly a matter of stratification and boundaries (for example, openness to the propertyless, the uneducated, women or immigrants) it is also a matter of how the public sphere incorporates and recognizes the diversity of identities that people bring to it from their manifold involvements in civil society. It is a matter of

DEVELOPING CAPABILITIES AND THE MANAGEMENT OF TRUST 141

whether in order to participate in such a public sphere, for example, women must act in ways previously characteristic of men and avoid addressing certain topics defined as appropriate to the private sphere. All attempts to render a single public discourse authoritative privilege certain topics, certain forms of speech and certain speakers.

(Calhoun 1996 in McKee 2004, p. 28)

Here, of course, we are talking about issues such as stratification and exclusion and of how the concerns of the marginalized or excluded come to be part of the public sphere. The idea of "counterpublics" is useful here. Counterpublics are groups of similar interest within which arguments can be developed to the point that they can invade and colonize the wider public sphere. Nancy Fraser provides an example from the feminist counterpublic, as follows:

Until quite recently feminists were in the minority in thinking that domestic violence against women was a matter of common concern and thus a legitimate topic of public discourse. The great majority of people considered this issue to be a private matter between what was presumed to be a fairly small number of heterosexual couples (and perhaps the social and legal professions who were supposed to deal with them). Then, feminists formed a subaltern counterpublic from which we disseminated a view of domestic violence as a widespread systematic feature of male dominated societies. Eventually, after sustained discursive contestation, we succeeded in making it a common concern.

(Fraser 2002 in McKee 2004, p. 163)

Education, as Freire (1970) pointed out so powerfully, is intimately involved in just such a process. In one of the few contemporary analyses of the intimate (but often unacknowledged) relationship between educational administration and social justice, Larson and Murtadha (2002) argue that while Freire's writing and work on educating oppressed populations has been used widely in curriculum theory, "leadership theorists have largely overlooked it. Nevertheless, Freire's arguments are as relevant to leadership as they are to teaching and learning"(2002, p. 146). In particular, Freire's analysis suggests that "many well-intentioned leaders maintain institutionalized inequity because they are committed to hierarchical logics that not only fail to question established norms but keep impoverished citizens out of decision making" (Larson and Murtadha 2002, p. 146) (see also Larson and Ovando 2001). The consequence is that:

Because poor and minority populations have learned to mistrust many public leaders, well-intentioned school leaders often have difficulty in earning their trust and cooperation. Freire explains that the lack of trust poor communities show to those who lead public institutions can be interpreted as an "inherent defect" in poor people, "evidence of their intrinsic deficiency". Since leaders need the cooperation of those they lead, educators can be tempted to resort to many of the same hierarchical and controlling practices used by dominant elites to oppress those they lead.

(Larson and Murtadha 2002, p. 147)

But such a withdrawal into a coercive form of leadership is not the only option for, as Freire argues, the role of leaders is "to consider seriously the reasons for mistrust on the part of oppressed populations, and to seek out true avenues of communion . . . helping the people to help themselves critically perceive the reality which oppresses them" (1970 in Larson and Murtadha 2002, p. 47). And indeed there seem to be plenty of oppressed people around for, as Amartya Sen suggests, we

live in a world with remarkable deprivation, destitution and oppression. There are many new problems as well as old ones, including the persistence of poverty and unfulfilled elementary needs, occurrence of famines and wide-spread hunger, violation of elementary political freedoms as well as of basic liberties, extensive areas of neglect of the interests and agency of women, and worsening threats to our environment and to the sustainability of our economic and social lives. Many of these deprivations can be observed, in one form or another, in rich countries as well as poor ones.

(Sen 1999, p. xi)

The Capability Approach

Sen's response to this situation is to insist on the intimate connection between individual agency (freedom) and social commitment:

Individual agency is, ultimately, central to addressing these deprivations. On the other hand, the freedom of agency that we individually have is inescapably qualified and constrained by the social, political and economic opportunities that are available to us. There is a deep complementarity between individual agency and social arrangements. It is important to give simultaneous recognition to the centrality of individual freedom *and* to the force of social influences on the extent and reach of individual freedom. To counter the problems that we face, we have to see individual freedom as a social commitment.

(Sen 1999, pp. xi–xii)

Social justice is then, in Sen's view, a matter of arranging our social commitments in ways that enhance individual freedom to live a valued life. In analyzing social justice, he argues, there is a strong case "for judging individual advantage in terms of the capabilities that a person has, that is, the substantive freedoms he or she enjoys to lead the kind of life that he or she has reason to value" (Sen 1999, p. 87).

As emphasized at the beginning of this chapter, capabilities, and the enhancement of individual capabilities through social arrangements, are at the heart of the issue of social justice, including management and leadership in education. Sen (1999, p. 10), offers five "types of freedom" or capabilities as being fundamental: (1) political freedoms, (2) economic facilities, (3) social opportunities, (4) transparency guarantees, and (5) protective security. Such capabilities are mutually supportive and are both ends and means of individual and social development. But while Sen gives a general outline (1999, pp. 38–40) and provides considerable empirical data in support of his analysis, he declines to give a specific content to these capabilities. This is a point of criticism by Martha Nussbaum, who argues that while

> I endorse these arguments . . . I think that they do not take us very far in thinking about social justice. They give us a general sense of what societies ought to be striving to achieve, but because of Sen's reluctance to make commitments about substance (which capabilities a society ought most centrally to pursue), even that guidance remains but an outline. And they give us no sense of what a minimum level of capability for a just society might be.
>
> (Nussbaum 2003, p. 35)

Nussbaum's solution is to give specific content to the capabilities advocated by Sen:

> The capabilities approach will supply definite and useful guidance . . . only if we formulate a definite list of the most central capabilities, even one that is tentative and revisable, using capabilities so defined to elaborate a partial account of social justice, a set of basic entitlements without which any society can lay claim to justice.
>
> (Nussbaum 2003, p. 36)

Moreover:

> These ten capabilities are supposed to be general goals that can be further specified by the society in question; as it works on the account of fundamental entitlements it wishes to endorse.
>
> (Nussbaum 2000, ch 1)
>
> But in some form all are part of a minimum account of social justice: a society that does not guarantee these to all its citizens, at some appropriate threshold level, falls short of being a fully just society, whatever its level of opulence. Moreover, the capabilities are held to be important for each and every person: each person is treated as an end and none as a mere adjunct or means to the ends of others.
>
> (Nussbaum 2003, p. 40)

Nussbaum's list of central human functional capabilities is (1) life, (2) bodily health, (3) bodily integrity, (4) senses, imagination, and thought, (5) emotions, (6) practical reason, (7) affiliation, (8) other species, (9) play, and (10) control over one's environment. Of these capabilities two are of particular importance:

> Practical Reason and Affiliation stand out as of special importance, since they both organize and suffuse all others, making their pursuit truly human. To take just one example, work, to be a truly human mode of functioning, must involve the availability of both practical reason and affiliation. It must involve being able to behave as a thinking being, not just a cog in a machine; and it must be capable of being done with and towards others in a way that involves mutual recognition of humanity.
>
> (Nussbaum 2000, p. 82)

Institutions, Capability Development and Social Justice

Whether one accepts Sen's broad definitions of capabilities or Nussbaum's elaborated list, the connection between capabilities—seen as the facilitating factors that allow individuals to live the kind of life they have reason to value—and the notion of social justice is clear. What is not immediately clear is how responsibility for the development of such capabilities can be allocated. Individuals must bear some of the responsibility, but so must institutions. Nussbaum argues that:

> One question that must certainly be confronted is the question of how to allocate the duties of promoting the capabilities, in a world that contains nations, economic agreements and agencies, other international agreements and agencies, corporations and individual people. To say that "we all" have the duties is all very well, and true. But it would be good if we could go further, saying at least something about the proper allocation of duties between individuals and institutions, and among institutions of various kinds.
>
> (Nussbaum 2002, p. 19)

Moreover, as institutions have capacities that individuals do not have, they must also take more responsibility for the support and development of capabilities:

> It is possible to argue cogently that institutions have both cognitive and causal powers that individuals do not have, powers that are pertinent to the allocation of responsibility. . . . [N]ations and corporations have powers of prediction and foresight that individuals in isolation do not have. It seems plausible that such facts give us a further reason to think of the responsibilities for promoting human capabilities as institutional.
>
> (Nussbaum 2002, p. 20)

Sen has a similar view:

> Individuals live and operate in a world of institutions. Our opportunities and prospects depend crucially on what institutions exist and how they function. Not only do institutions contribute to our freedoms, their roles can be sensibly evaluated in the light of their contributions to our freedom. To see development as freedom provides a perspective in which institutional assessment can systematically occur.
>
> (1999, p. 142)

This brings us back to the issues of administration and leadership, specifically educational administration. If the above argument holds then there is an intimate relationship between social justice (seen as the equitable promotion of human capabilities) and the administration of institutions. Institutions can be held to account for their contribution to (or denial of) social justice in terms of their contribution to the development and extension of human capabilities. This is a somewhat different and more complex notion of accountability than that (rather perverse) form outlined by O'Neill. It is also particularly apposite for the calling of education to account.

Educational Administration and Leadership, and Capabilities

Rather than the narrow and distorting forms of accountability currently demanded through standardized tests and performance audits, the capabilities perspective demands that the role of education in promoting the full range of human capabilities be considered and acted upon. This is both a curricular responsibility and a pedagogical one: curricular, in the sense that the drive toward fuller development of human capabilities requires a curriculum appropriately shaped toward achieving such ends in particular circumstances; pedagogical, in the sense that the circumstances that students bring to the educational context must be taken into account in the teaching relationship. Larson and Murtadha (2000) capture these requirements rather well in their description of the person education should be aiming to encourage. Such a person, they say,

> should be capable of practical reason, being able to form a conception of the good life and to engage in critical reflection about planning one's life. This also means being able to recognize and live with concern for other human beings, to engage in various forms of social interaction, to laugh, to play, to enjoy recreational activities, to imagine the situation of another and to have compassion for that person, and to have the capacity for both justice and friendship.
>
> (2002, p. 155)

Such accomplishments are not notably the subject of high stakes testing! And, of course, such concerns have traditionally been excluded from the province of educational administration through the separation of departments of "Leadership" from departments of curriculum and instruction. But surely leadership cannot be divorced from an appropriate conception of how a "good education" is to be achieved through the mechanisms of curriculum and instruction. And a good education in terms of the above argument must be intimately related to ideas of social justice. This being so, the whole condition of students' lives is an essential consideration, not simply their academic performance:

> For educational leaders, a focus on capabilities as worthy educational goals necessitates promoting a greater measure of equality than exists among most schools struggling with the legacies of racism, sexism and classism. This approach suggests that if children receive educational and material support, they can become fully capable of human action and expression. Freedom from violence, unconditional support, and concern for health and nutrition are educational considerations beyond today's hollow and entirely insufficient demands for improving academic achievement.
>
> (Larson and Murtadha 2002, p. 155)

To be fair to the field of educational administration, more than one voice is currently being raised with such goals in sight. Murphy (2002), for instance, argues that principals must become "moral stewards" whose actions are "anchored in issues such as justice, community and schools that function for all children and youth" (2002, p. 75). Just what constitutes social justice is, however, left rather open.

More powerfully, Furman and Starratt (2002) link the pursuit of social justice through education with the reinvigoration of democratic community. Drawing on Apple and Beane (1995), they argue that the central concerns of democratic, socially just schools include:

> The open flow of ideas, regardless of their popularity, that enables people to be as fully informed as possible;
> The use of critical reflection and analysis to evaluate ideas, problems and policies;
> Concern for others and the "common good";
> Concern for the dignity and rights of individuals and minorities.
>
> (Furman and Starratt 2002, p. 106)

Starratt works this thesis out more fully in his book, *Centering Educational Administration* (2003), concluding that educational administrators have a vastly more responsible role than that allocated to them by functionalist accounts of educational administration that limit them to the management of efficiency, accountability, performance, and control. For Starratt educational administrators:

> are called upon to assume the role of citizens taking responsibility to work together to make our world a better place: more human, more just, more civil, and more in harmony with our natural environment. In other words, we are accountable to both our ancestors who struggled to create the world we live in and to our progeny who must live with the public choices we collectively make. Thus, our accountability is not simply a legal concern, not simply an academic concern, not, indeed, simply a social concern to protect children. Our accountability is also a moral concern to bring the work of learning to bear on our collective responsibilities.
>
> (Starratt 2003, pp. 228–229)

Bringing the work of learning to bear on our collective responsibilities is surely the preeminent task of educators and educational administrators. Seen within the context of a capability approach, such a charge might go far to redress the damage done by decades of educational administration informed by a woefully inadequate understanding of both education and administration and of their necessary links to the pursuit of social justice.

But what might such learning look like? Walker (2006) is one the few theorists to have attempted to draw out the implications of the capability approach for education. Arguing that "the capability approach focuses on people's own reflective, informed choice of ways of living that they deem important and valuable, and the self-determination of ends and values in life" (p. 21), Walker goes on to suggest that in higher education at least this approach "foregrounds human development, agency, well-being and freedom"(p. 42).

Capabilities, Learning, and Curriculum

Such a view is in keeping with other contemporary theorists of education such as Starratt in his call for an emphasis on the "three qualities of a fully human person: . . . autonomy, connectedness and transcendence" (2003, p. 137) as the essential foci of schooling. Starratt's view places autonomy, agency, and freedom within a context of community while simultaneously invoking the idea of transcendence of both self and community:

> Autonomy makes sense only in relation to other autonomous persons, when the uniqueness and wealth of each person can be mutually appreciated and celebrated. Connectedness means that one is connected to someone or something other than oneself. Hence, it requires an empathetic embrace of what is different from the autonomous actor to make and sustain the connection.

Community enables the autonomous individual to belong to something larger; it gives the individual roots in both past and present. However, community is not automatically self-sustaining. It is sustained by autonomous individuals who transcend self-interest to promote the common good.

(2003, p. 139)

In fact, Starratt has introduced a fourth idea here. In addition to autonomy, connectedness, and transcendence, his notion of the embrace of *difference* is an essential component of a capabilities-based curriculum. It is this idea that enlarges the concept of community from the localized and exclusionary "virtuous community" outlined by Sergiovanni (1994) to one that "transcends national boundaries, economic competition, and the desire for ever more sophisticated technology" (Starratt 2003 p. 64). Moreover, it is such difference and the search for meaning in such difference that Starratt sees as spur to learning.

This is essentially a cultural process, but not one seen simply as an induction into a single way of life. Rather, as I have suggested (Bates 2005, p. 105), it is one based on a fundamental premise that in terms of both self and community the foundational value is "the freedom to build a personal life which respects an equal freedom for others to do likewise." This is a position very close to Sen's argument concerning "development as freedom" (1999).

Elsewhere (Bates 2006) I have elaborated the consequences of this principle in terms of Sen's "global curriculum," a curriculum that must be based around understanding rather than simply a description of the Other: of other cultures and their ways of life, values, artifacts, and histories. This will mean bringing into the curriculum ways of understanding that allow us to not simply empathize with but also comprehend the nature of subjugated knowledges—of those understandings that are pushed to the periphery of our comprehension by our xenophobia. It means a curriculum that is not narrowed by a prism that reduces the complex to the simple, the diverse to the unitary, and imposes simplistic judgments of good and evil, defining the Other as having a nature irredeemably different from our own. Moreover, such a curriculum would take seriously "real and observable inequalities as its starting point . . . [making] an active attempt to compensate for them" (Touraine 2000, p. 270; see also Bates 2006).

Such an agenda is avowedly cultural in its orientation in that it seeks the building of capabilities for personal freedom and the realization of such freedom within complex, differentiated, and changing cultural environments. Indeed, as Sen and Nussbaum both suggest, the building of such capabilities for freedom is as much a matter of building the capacity to challenge *cultural* hegemony as it is of developing the capacity to challenge *economic* hegemony.

The implications for curriculum and pedagogy are profound. Rather than "values education" being the inculcation of particular values derived from a specific cultural tradition, this means "in reality, that all ways of life are continuously held up to scrutiny and evaluation. Moreover, living alongside each other allows for the process of cultural hybridization that Rizvi (1997) among others, points to as a certain consequence of our increasing proximity. Touraine, in his examination of the question, "Can we live together?" also argues that "cultures are not, at least in the modern world in which we live, separate and self-contained entities, but modes of managing change as well as systems of order" (2000, p. 177).

Learning is, then, very much a matter of coming to terms with cultures and differences as ways of understanding our world and those of others, and of using these as "sources of the self" (Taylor 1989) in the construction of our identities and in the recreation of our cultures. This, however, is to take a moral-political perspective on the role of education in developing capabilities. It is a perspective that envisages the development of a democratic learning society founded on the principles of liberty, equality, justice, and democracy (Bocock 1986; Carr and Hartnett 1996; Quicke 1999). Fundamental to education in such a society would be a twin commitment to its being *person centered* and *community oriented* (Quicke 1999, p. 3). Within such an education system:

> The learner . . . is constituted as a person—someone who actually or potentially has the capacity to make moral choices, act autonomously and think rationally—and learning is about the development of persons as unique individuals through being active participants in democratic learning communities. Such participation empowers them to act upon the world around them and transform it, and to act upon themselves in the same way.
>
> (Quicke 1999, p. 3)

Quicke here echoes Carr and Hartnett in their defense of the principles of classical democracy and of a resulting democratic education:

> In this version of democracy, the emphasis is on democracy as a "way of life" where individuals realize their potentialities through active participation in the life of their society, and [a] "democratic society is thus an educative society [a "learning society"] whose citizens enjoy equal opportunities for self-development, self-fulfillment and self-determination."
>
> (Carr and Hartnett 1996 in Quicke 1999, p. 3)

Such a moral-political view of democracy and the ideal of a democratic learning society is contrasted with the instrumental approach to education increasingly incorporated into modern education systems where modernity and the "administered world" (Starratt 2003), combined with the stresses of "turbo capitalism" (Luttwak 1995) and globalization define the essentials of learning as the (temporary) mastery of an ever-changing repertoire of skills under conditions of risk, uncertainty, and competition. Markets take the place of community in such a society where continuous strategic action frustrates the development of self and community alike (Bates 2006; Touraine 2000). The result is problematic at both social and personal levels as the social relations of production are de-socialized and the relationship between production, social system and personality collapses. The dark side of modernity is, then, a dissociation of personality from system, sociability from production and the dissolving of culture in the solvent of the strategic risks of market 'society.'

The alternative is, as Ranson and Stewart (1998) suggest, a learning society. This is:

> a different form of society, one that places active citizenship as the organizing principle of learning. By creating conditions for a new participatory democracy, citizens will be enabled to find themselves, individually and together, through making the common wealth of their communities.
>
> (Ranson and Stewart 1998, p. 254)

In such a society, they argue, several substantive characteristics will be present. First, the society will see education as a process that enables *collective change* through learning about society itself, how it is changing, how it needs to change its processes of learning. Second, it will encourage individual *agency* through the active participation of citizens in the making and remaking of their communities. Third, through *dialogue* it will encourage a conversation about value differences and enable them to be subjected to critical scrutiny. Fourth, it will be concerned with creating democratic institutional forms through which active citizenship can be exercised (Ranson and Stewart 1998, p. 254).

Bottery (2004) elaborated a similar view of the learning society, showing how it needs to be underpinned by certain key educational objectives. In summary these are as follows:

- An *economic productivity* objective through which individuals can develop the capabilities required to earn a living and contribute to economic wealth
- A *democratic* objective through which individuals can acquire the capabilities required to contribute to the development of a democratic state
- A *welfare state* objective through which the capabilities required to redress inequalities and contribute to the social and political community can be acquired

- An *interpersonal skills* objective through which the capabilities of living together in a fulfilling manner can be developed
- A *social values* objective through which the values of equity, care, harmony, environmental concern, and democratic orientation can be encouraged
- An *epistemological objective* concerned with developing the capability for deep understanding and the possibilities and limitations of various kinds of knowledge
- A *personal development* objective through which an appropriate sense of self and identity can be formed
- An *environmental* objective through which the capability to understand the interdependence of species and human impact upon the planet can be developed

A similar set of principles for a curriculum for "new times" is identified by Quicke (1999) where he sees six themes as of crucial importance. First, one of the most critical tasks in our pluralistic, constantly changing world, is that of forming a *self-identity*. While in traditional societies fairly stable identities were formed from a relatively restricted range of "sources of the self" (Taylor 1989), "new times" offer a multiplicity of choices and therefore require "a more explicit form of self-understanding which enables people to think reflexively, critically and imaginatively about their life projects and *self identity*" (Quicke 1999, p. 7). Second, as self identity is formed through interaction with others, the capacity for *collaboration* is absolutely crucial: "a commitment to collaborative, democratic relationships is of great value to the self because it anticipates social arrangements in the wider community which ensure space for autonomy-enhancing practices" (Quicke 1999, p. 7).

Third, as one of the first structures through which identity and collaborative practices are formed, the *family* is important as a model of the ways in which such practices change and evolve through negotiation and adaptation to a changing environment since "even the 'structures of intimacy' amongst persons in 'close' relationships are open to revision and reconstruction in the light of changing perceptions of the self and the social world" (Quicke 1999, p. 8). Fourth, *cultural pluralism* is one of the key features of the contemporary social world and while such increased pluralism gives rise to potential tensions between groups, it also gives rise to a range of possible options while encouraging cultural change so that as "the scope for cultural interaction increases in a multiplicity of contexts via new communications systems, so the opportunities for experimentation and negotiation give rise to an increasing diversity of new cultural forms" (Quicke 1999, p. 9).

Fifth, *work and economic life* is a central source of identity as well as wealth creation. Here the changing forms of production and consumption, the shortening of the "shelf-life" of goods and services as well as employment opportunities produce both opportunities for the exercise of agency and considerable instability and risk. One of the results of these changes has been that "[p]revious forms of monitoring of workers are displaced by self-monitoring, and the systems of rules and resources which constitute the means of production also become the object of reflection by agency" (Quicke 1999, p. 9). The positive aspect of this change is that "new" workers "need to be adaptable and flexible, but also capable of seizing opportunities to develop their skills, to be creative and to take responsibility for their own work in a way that maximizes the scope for self-regulation and autonomy" (Quicke 1999, p. 10). Finally, the role of knowledge, particularly scientific knowledge, has become crucial in the learning society. But as it has become more central it has also become a source of risk as well as a source of "progress." The theme of *science in a risk society* therefore becomes a central theme of the curriculum, enabling development of learners' "capacity to evaluate expertise and think rationally about appropriate courses of action in conditions of risk and uncertainty" (Quicke 1999, p. 11).

There is, then, a substantial body of theory that advocates an educational program quite consistent with the capability approach outlined by Sen and Nussbaum. It is, of course, a theoretical tradition that goes back to Dewey and his insistence that knowing comes about through doing, through active participation in production, and active involvement in democratic social processes; in short, education and agency are inextricably linked, as are agency, production, and citizenship. The curriculum for what we now call the learning society would inevitably, therefore, include:

> instruction in the historic background of present conditions; training in science to give intelligence and initiative in dealing with material and agencies of production; and the study of economics, civics and politics to bring the future worker in touch with the problems of the day. . . . Above all, it would train the power of readaptation to changing conditions so that future workers would not be become blindly subject to a fate imposed upon them.
> (Dewey 1933, p. 372)

Dewey, like Sen, saw agency as central to freedom and saw production and participation as fundamental to well-being. Walker's summary of the capability approach could well be a summary of Dewey's fundamental principles, when she explains that the capability approach "foregrounds human development, agency, well-being and freedom. It offers a compelling counterweight to neo-liberal human capital interpretations and

practices of . . . education as only for economic productivity and employment" (Walker 2006, p. 42).

Pedagogy

The question that arises from such analyses is, therefore, what kind of
pedagogy might enable such a curriculum? Lingard et al. (2003), following on from the work of Newman and Associates (1996) have developed
an analysis of pedagogical practices that allows for the development of
such a pedagogy. Called "Productive Pedagogies," this approach to learning focuses attention on four main categories of interaction in learning
environments. These are as follows:

- The *intellectual quality* of the interactions, which includes the treatment of knowledge as problematic and provisional, the depth of
  knowledge developed through interaction, the depth of a student's
  understanding, the substantive nature of the conversation, the development of higher order thinking, and the use of metalanguages
- The *connectedness* of classroom activity to the world beyond the
  classroom, the integration of various forms of knowledge, the links
  to background knowledge, and the development of a problem-based
  curriculum linked to the participants' backgrounds
- The *supportiveness* of the classroom environment including student
  direction, explicit performance criteria, social support, academic
  engagement, and student self-direction
- The *engagement with difference* around knowledge of other cultures,
  active citizenship, narrative appreciation, group identity and the
  development of learning communities, and the representation of
  difference in those communities

On the basis of such pedagogy, Lingard et al. argue that leadership in the
classroom, school, or learning organization can be reconfigured in ways
that promote the objectives of the kind of learning society discussed
above. Such leadership would no longer center around the implementation of directives regarding a curriculum based solely on skills and com-
pliance but be engaged actively in the emancipatory possibilities of
learning. They argue for an approach in which

> productive leadership encourages intellectual debates and discussions about
> the purposes, nature and content of a quality education; promotes critical
> reflections upon practices; sponsors action research within the school; and
> seeks to ensure that this intellectual work connects with the concerns of
> teachers, students, parents and the broader educational community. Such

leadership also ensures that teachers, and others working within schools, are provided with the support and structures necessary to engage in intellectual discussions about their work, to reflect on the reform processes within their schools, as well as their pedagogical and assessment practices.

(Lingard et al. 2003, p. 20)

Moreover, such practices need to be rooted in concerns for social justice and an acknowledgment that as traditionally organized, most schools have tended to perpetuate rather than challenge inequities:

[P]roductive leadership also demonstrates a concern with social outcomes. Because schools have traditionally served the interests of dominant cultures, in curriculum, pedagogical and assessment terms, productive leadership works towards ameliorating inequalities amongst the school community. Our normative position is that educational leaders need to have an awareness of social justice. We believe that these issues are equally important in schools servicing dominant groups as they are in schools serving marginalized groups. Recognizing that schools have been traditionally hierarchical institutions where teachers (and students) have had little to say in the running of the school, we support a democratic vision of school organization.

(Lingard et al. 2003, pp. 20–21)

Such a position echoes that of educators such as Apple and Beane (1995) and Mills (1997) as well as the call of Darling-Hammond (1997) for a paradigm shift in the ways in which schools are organized and led:

[There needs to be] . . . a paradigm shift in how we think about the management and purpose of schools: from hierarchical, factory model institutions where teachers treated as semi-skilled assembly line workers, process students for their slots in society, to professional communities where student success is supported by the collaborative efforts of knowledgeable teachers who are organized to address the needs of diverse learners.

(Darling-Hammond 1997, p. 332)

In Conclusion

Which brings me back to the starting point of this chapter. While educational leadership has traditionally been conceived of as the administration of curricular, pedagogical, and assessment practices devised elsewhere, a fully professionalized form of educational leadership would be based on educational rather than administrative principles. Moreover, these educational principles would themselves be based upon a conception of the learning society that took the development of capabilities centered around ideas of human development, agency, well-being, and freedom as central, thus claiming that the development of a truly democratic and free society should be the purpose behind human activity, one to which the economic development of such societies should be directed. A capabilities approach to educational administration may well enable the transformation of the field into one of genuine educational leadership for a just society.

References

Apple, Michael, and James Beane, eds. 1995. *Democratic schools.* Alexandria, VA: Association for Supervision and Curriculum Development.

Ball, Stephen. 1994. *Education reform: A critical post-structuralist approach.* Buckingham: Open University Press.

Bates, Richard J. 1980. Educational administration, the sociology of science and the management of knowledge. *Educational Administration Quarterly* 16: 1–20.

———. 1983. Educational administration and the management of knowledge. Geelong, Australia: Deakin University Press.

———. 2005. Towards a global curriculum. *Arts and Humanities in Higher Education* 4:95–109.

———. 2006. Educational administration and social justice. *Education, Citizenship and Social Justice* 1:171–187.

Bocock, R. 1986. *Hegemony.* London: Tavistock.

Bottery, M. 2004. *The challenges of educational leadership.* London: Paul Chapman Publishing.

Boyan, Norman J., ed. 1988. *Handbook of research on educational administration.* New York: Longman.

Calhoun, Craig. 1996. Social theory and the public sphere. In *The Blackwell companion to social theory,* edited by B. S. Turner. Oxford: Blackwell.

Callahan, Raymond E. 1962. *Education and the cult of efficiency.* Chicago: Chicago University Press.

Carr, Wilfrid, and Anthony Hartnett. 1996. *Education and the struggle for democracy.* Buckingham: Open University Press.

Darling-Hammond, Linda. 1997. The right to learn: A blueprint for creating schools that work. San Francisco: Jossey-Bass.

Dewey, John. 1933. *Democracy and education.* New York: Macmillan. (First published in 1916.)

Foster, William. 1986. *Paradigms and promises.* Buffalo: Prometheus Books.

Fraser, Nancy. 2002. Recognition without ethics? In *Recognition and difference: Politics, identity, multiculture,* edited by S. Lash and M. Featherstone. London: Sage.

Freire, Paulo. 1970. *Pedagogy of the oppressed.* New York: Seabury.

Furman, Gail, and Robert Starratt. 2002. Leadership for democratic community in schools. In *The educational leadership challenge: Redefining leadership for the 21st century,* edited by J. Murphy. Chicago: Chicago University Press.

Greenfield, T. Barr. 1975. Theory about organization: A new perspective and its implications for schools. In *Administering education: International challenges,* edited by M. Hughes. London: Athlone Press.

Larson, Colleen, and Khaula Murtadha. 2002 leadership for social justice. In Murphy, *The educational leadership challenge: Redefining leadership for the 21st century,* edited by J. Murphy. Chicago: Chicago University Press.

Larson, Colleen, and Carlos Ovando. 2001. The color of bureaucracy: The politics of equity in multicultural school communities. Belmont, CA: Wadsworth.

Lingard, B., D. Hayes, M. Mills, and P. Christie. 2003. *Leading learning: Making hope practical in schools.* Buckingham: Open University Press.

Luttwak, Edward. 1995. Turbo charged capitalism and its consequences. *London Review of Books,* November 22: 6–7.

McKee, Alan. 2004. *The public sphere: An introduction.* Cambridge: Cambridge University Press.

Mills, Martin. 1997. Towards a disruptive pedagogy: Creating spaces for student and teacher resistance to injustice. *International Studies in the Sociology of Education* 17:315–326.

Murphy, Joseph. 2002. Reculturing the profession of educational leadership. In *The educational leadership challenge: Redefining leadership for the 21st century,* edited by J. Murphy. Chicago: Chicago University Press.

Newmann, Fred M., and Associates. 1996. *Authentic achievement: Restructuring schools for intellectual quality.* San Francisco: Jossey-Bass.

Nussbaum, Martha C. 2000. *Women and human development: The capabilities approach.* Cambridge: Cambridge University Press.

———. 2002. Justice for the extended world: Capabilities across national boundaries. Tanner Lecture, University of Canberra.

———. 2003. Capabilities as fundamental entitlements: Sen and social justice. *Feminist Economics* 9 (2–3): 33–59.

O'Neill, Onora. 2002. *A question of trust.* Cambridge: Cambridge University Press.

Olssen, Mark, John Codd, and Anne-Marie O'Neill. 2004. *Education policy: Globalization, citizenship and democracy.* London: Sage.

Quicke, John. 1999. *A curriculum for life.* Buckingham: Open University Press.

Ranson, Stewart, and J. Stewart. 1998. The learning democracy. In *Inside the learning society,* edited by S. Ranson. London: Cassell.

Rizvi, Fazil. 1997. Beyond the East-West divide: Education and the dynamics of Australia-Asia relations. *Australian Educational Researcher* 24:13–26.

Sen, Amartya. 1999. *Development as Freedom.* Oxford: Oxford University Press.

Sergiovanni, Thomas J. 1994. *Building community in schools.* San Francisco: Jossey-Bass.

Starratt, Robert. 2003. *Centering educational administration.* London: Lawrence Erlbaum.

Taylor, Charles. 1989. *Sources of the self.* Cambridge, MA: Harvard University Press.

Taylor, Frederick W. 1911. *The principles of scientific management.* New York: Harper.

Touraine, Alain. 2000. *Can we live together?* Stanford, CA: Stanford University Press.

Walker, Melanie. 2006. *Higher education pedagogies.* Maidenhead, UK: Open University Press and McGraw-Hill Education.

Wise, Arthur E. 1979. *Legislated learning: The bureaucratization of the American classroom.* Berkeley: University of California Press.

8

Education and Capabilities in Bangladesh

Janet Raynor

Overview

Based on Martha Nussbaum's list of central human functional capabilities, this chapter is an empirical study of the links between education and the development of girls' and women's capabilities in Bangladesh. The study arises from the expansion of girls' education in Bangladesh; the widespread assumption that education helps lead to empowerment (which I take to be the development of capabilities); the claim by some that Bangladesh has already achieved the Millennium Development Goal (MDG) of the empowerment of women, and the evidence indicating otherwise. Through interviews with adolescent girls about what going to school and "being educated" means to them, there is an exploration of which capabilities are enhanced by the educational process in Bangladesh, and which are not. I argue that education cannot be regarded as a basic capability unless it specifically addresses the process of developing the capabilities necessary to live a life one has reason to value.

Achieving the MDGs

At the United Nations World Summit held in New York in September 2005, leaders of the nations of the world met to review progress toward the MDGs agreed in 2000, most of which were set to be met by 2015. Of particular interest was MDG 3, to "promote gender equality and empower women," which was the only goal to have an interim target for 2005.

That target was to "eliminate gender disparity in primary and secondary education preferably by 2005 and in all levels of education no later than 2015" (United Nations 2000). Out of this summit came reports that Bangladesh had already achieved MDG 3; this report was picked up and circulated around the world (e.g. News from Bangladesh 2005; Zia 2005; Arabic News 2005; Asia-Pacific Daily News Review 2005). Unfortunately, although Bangladesh can be seen to be moving toward equality and empowerment in some areas, the claim is quite simply not true. That such a claim was made, unquestioningly published, and then repeated in various forms indicates a widespread misunderstanding of this particular MDG, and of the concepts or measurement of equality and empowerment. That the internationally agreed target for reaching the goal is to eliminate gender disparities in education indicates an enormous confidence in—or naïveté about—the power of education to bring about social justice. In this case, the social justice under discussion is that of women's empowerment, or capabilities, and the role that education might have in enhancing it.

In this chapter, the aim is, therefore, to examine girls' educational experiences in Bangladesh through a capabilities lens to try to get a better picture of how close Bangladesh is to achieving MDG 3, or even the target of eliminating gender disparities in education. The chapter reviews the expansion of girls' education in Bangladesh, the widespread claim that such education helps lead to empowerment, and the limited evidence indicating that it might not. Many of the "empowerment" claims or reports work with loose or undefined conceptions of either education or empowerment, and in Bangladesh there have been few attempts to measure specific education-linked empowerment other than the proxy indicator of numbers of girls in school. Nussbaum's list of central human functional capabilities (2000, 2002) can be seen as one approach to encapsulating the elements that make up empowerment, and here I use her list as a framework for identifying links between girls' education or schooling and the development of their capabilities, or what could be called empowerment. The analysis rests on data collected through an interview-based study of adolescent girls in rural Bangladesh.

Links between the capability approach and education are clearly defined, but the term *capabilities*—as used by Amartya Sen—has rarely been used in connection with education policy and practice in Bangladesh. This may be partly because there is no easy shorthand term for the process of having a person's capabilities developed so that they are able to live a life that they have reason to value, or for the state of having had those capabilities developed. Therefore I am appropriating the more commonly used terms of *empowerment* and *empowered* for this purpose.

MDG 3 and Girls' Education in Bangladesh

In terms of enrollment, there have been impressive quantitative advances in girls' education in Bangladesh since liberation in 1971. At that time, female literacy was only about 11 percent. The constitution drawn up in 1972 guaranteed equal rights to all, and stressed government commitment to free and compulsory basic education, made law in 1990; by 2002, gender parity of enrollment had been achieved (Ahammed 2003) for at least the first eight years of schooling. The intention here is not to diminish these considerable achievements, but to try to generate a more critical account of what this education means for girls' lives. To go back to the claim that Bangladesh had already achieved MDG 3: in the statement prepared for the prime minister to present at the World Summit, it was reported that Bangladesh had achieved the MDG of "eliminating gender disparity in primary and secondary schools" (Zia 2005). The claim was based on statistics showing that there was approximate parity of enrollment in primary and earlier secondary years in Bangladesh. It should be noted that the claims relate to the *target* of eliminating gender disparity in education rather than the *goal* of promoting gender equality and empowering women. Therefore, even if the target has been reached, the goal has not. In addition, Bangladesh once again scored extremely poorly on UNDP's overall Gender Empowerment Measure, ranking 79th out of the 80 countries included (UNDP 2005). Bangladesh, therefore, still has some way to go before it reaches the goal of equality and empowerment. This fact is acknowledged in the Bangladesh MDG report, which recognizes that real empowerment is "still a distant goal," and must include not only education, but also progress in areas such as health, violence against women, and economic and political disparities.

The MDG 3 target of gender parity in education agreed in 2000 is simply an indicator of progress toward the goal of empowerment. Measurement of gender parity in education does not have to be restricted to enrollment; it could, for example, relate to parity of achievement, outcomes, curricular content, or educational processes. And measurement of empowerment does not have to be restricted to education, as can be seen from the much more strongly worded and expanded statement on the promotion of gender equality in the resolution adopted at the 2005 World Summit (United Nations 2005). This strengthening of commitment came about largely as a result of women's groups lobbying at the summit, and there are now seven agreed target areas, of which education is one. Others include rights of ownership; access to reproductive health, labor markets, assets, and resources; a recommitment to eliminating discrimination and violence against women; and increased representation and participation of

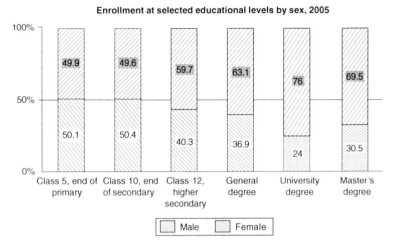

Figure 8.1   Gender inequalities in enrollment in Bangladesh

Source: Data extracted from BANBEIS 2006.

women in government decision-making bodies (Center for Women's Global Leadership 2005). My interest is in what ways education can help bring about change in all such areas.

However, the fact that enrollment has been set as one of the MDG indicators leads to reporting only on this very narrow measure of gender parity. Official statistics for Bangladesh in 2005 (those included in figure 8.1 have been extracted from BANBEIS 2006) indicate that girls' and boys' enrollment is about on a par at primary and early secondary level, but with a significant gap opening up by higher secondary school, where girls are only 40 percent of the school population. The gaps widen further at tertiary level, with only about a quarter of university students being women. There are even wider gender gaps in the Islamic schools and colleges and in specialist technical and vocational education schools. Even if we consider only parity of enrolment, because these gaps exist, Bangladesh cannot yet claim to have reached the 2005 target of eliminating gender disparity in primary and secondary education, and it is highly unlikely that it will meet the 2015 target of gender parity in all levels of education.

But enrollment is only one part of gender parity. Another area where progress toward parity is relatively easily measured is examination results. For example, the scholarship examination taken at the end of primary school is the first national examination a child might face, and those who are awarded a scholarship receive free tuition at secondary level. In theory, the examination is open to all schoolchildren; in practice, only the most privileged are entered. The most recent year for which I have been able to find figures is 2001 (DPE 2002). These suggest that girls were only 44.8 percent of the total entries, and only 41.15 percent of the total number of children who passed. That is, girls are less likely to be entered for the examination, and less likely to pass. Those combined disparities lead to a gap of nearly 18 percent between the 41.15 percent of successful candidates who were girls, and the 58.5 percent who were boys.

Moreover, such disparities start early, and are evident throughout the education system. There are similar gaps for the Secondary School Certificate examination (roughly equivalent to GCSE), taken at the end of Class 10. The gender gap was 12 percent in 2005 (based on figures published in *New Nation* 2005). Of the total number of children who passed the examination, 44 percent were girls and 56 percent boys. There are similar gender disparities in terms of subject choice, where, for example, boys are twice as likely as girls to be included in the high-status science stream, and even bigger gaps at the tertiary level (BANBEIS 2005). Such gaps are largely ignored. As Robeyns (2006) suggests, in a rights-based approach, governments can be satisfied with equal enrollment even where there are significant gender inequalities in outcomes.

There are many possible reasons for the disparities. Given that it is now almost universally accepted that girls are at least as able as boys, there is a need to know more about the gaps, and what can be seen as a capabilities failure. For education to help move toward equality and empowerment, we need to identify in what ways the education system helps or hinders the development of girls' capabilities.

Girls' Education through a Capabilities Lens

Gender and education in Bangladesh are seen in terms of enrollment, or the instrumentalist objectives of girls' and women's education rather than the freedoms that education can foster. The central question in the capability approach is "What is this person able to do and to be?" In relation to education, Walker argues that the value of the capability approach is that it "enables us to ask a different set of questions about education . . . about what education enables us to do and to be" (Walker 2006, p. 164). In this chapter, influenced by Nussbaum's feminist perspective on capabilities, the question is, "How does education in Bangladesh help to enhance the doings and beings of girls, and the women they could become?"

A basic education, which in Bangladesh is generally seen to be five years of schooling of questionable quality, is not enough to help girls and women achieve the capabilities defined by Nussbaum, or the "empowerment" widely claimed. Therefore, my focus is on adolescent girls, most of whom have gone beyond what can officially be viewed as "basic education." But I am also looking at education in the wider sense of a system or set of systems—formal, nonformal, or informal—that provide people with the skills, knowledge, and support they need to enable them to develop their capabilities to the fullest possible extent. As Nussbaum argues, "Human beings are creatures such that, provided with *the right educational and material support*, they can become fully capable of the human functions" (2002, p. 62, my emphasis). What constitutes "right educational support" needs further explication, and for this we need to look beyond enrollment and retention figures.

Writing on women's capabilities and social justice, Nussbaum (2002) is critical of traditional approaches to development and ways of measuring human capabilities, and her general critique of such approaches is applicable to approaches to education in Bangladesh. For example, economic approaches focus on the amount of money available, but Nussbaum insists that we look at people's lives. Bangladesh's much-lauded female stipend program is essentially a matter of paying parents to send their girls to school; there has been little concern about whether the schooling they receive is transformative. Nussbaum sees the problem with resource-based approaches being the variation in people's ability to convert resources into functionings, and that such approaches often substitute something easily measured for what should be measured. An example of this can be seen in Bangladesh with the flurry of building school toilets for girls—but no one checking whether the girls are using them, or why adolescent girls are absent from school for up to one week a month. As for preference-based approaches, Nussbaum notes that women in particular have often been deprived of (among other things) the education and information they need to be aware of the full range of options before deciding on preferences. Simply being at school is no guarantee of being given what is needed to be able to identify individual preferences, and talking to girls in Bangladesh shows just how limited a set of preferences is offered to them.

Unterhalter (2003) argues that education has been undertheorized, and criticizes Sen, who regards education as a fundamental or basic capability, but who fails to distinguish between education and schooling. Sen's view, she says, "does not take account of education facilities and processes that might not enhance freedom" and stresses that school education in particular is not necessarily an enlargement of freedom. She gives as examples "village schools in some states in India where girls might learn subordination, because they are forced to sit at the back of the classrooms in the dark, or because of the ways in which they are portrayed in textbooks" (Unterhalter 2003, pp 10–11). She makes a clear distinction between "schooling" and "education," and the point that gender equity in education initiatives "rest on the assumption that access to education is the key to the social and economic well-being of girls and women . . . factors such as content and pedagogy, and issues about how education changes subjectivities are ignored" (Unterhalter 1999, p. 51). Schooling may not bring about an enhancement of capabilities so that "formal schooling may as much be a case of capability deprivation as of human capability in development" (Unterhalter 2003, p. 8).

Similar patterns are to be seen in Bangladesh, and are part of the informal education context in school. Jeffery and Jeffery (1998), writing about South Asia, reviewed sweeping claims for the power of education, and insist that there is no causal link between education and women's autonomy. Arends-Kuenning and Amin (2001), writing on education in Bangladesh using a capabilities analysis, assert that education as it stands does not change or challenge patriarchy or have any transformatory significance, and insist that the content of education must change. This is also reflected in my own experience of Bangladesh: education's potential to empower is not being harnessed. For example, in a series of 30 observations of mixed classes in secondary and tertiary classrooms in Bangladesh, a simple count of number and duration of utterances indicated a consistent bias toward male students. Visual observations showed that girls and women had less control of classroom resources; were less likely to make eye contact with the teacher; less likely to raise their hands to answer questions; and if they did raise their hand, it was rarely above shoulder level. That is, they had learned to remain unobtrusive in class; the males had learned few such inhibitions (Raynor and Pervin 2005).

Nussbaum argues that the capability approach helps us to construct a normative conception of social justice, and that such an approach is well suited to focus on gender issues. However, she says, this can best be done by specifying a definite set of capabilities, highlighting the ones we see as the most important to protect (2003). To this end, she has developed a list that she sees as determining a decent social minimum, and it is this list that I am using as a reference point. Education is included in this list of central human functionings, most obviously in two of the ten capabilities specified: *senses, imagination, and thought,* and *practical reason.* But the development of other capabilities could be—or should be—part of an education in its broadest sense. As Nussbaum says, education should include, but not be limited to, literacy, and basic mathematics and science training. So, for example, education that developed the capabilities of *affiliation* and *bodily integrity* would ensure a space for girls to appear without shame in a public setting such as a school, to raise their eyes and their hands, and to participate equally in class; this would also prepare them for taking on public roles later in life. Such an education would also teach boys and teachers to respect that space.

While there may be some concerns that such an education could threaten cultural diversity, it is hard to see how education can be the means of achieving MDG 3 without embracing concepts such as Nussbaum's explicit and comprehensive—and to her irreducible—set of capabilities. Other writers have developed lists of capabilities specific to particular contexts, and in consultation with specific groups—for example, conceptualizing gender inequality in Western societies (Robeyns 2003), gender equity in education in South Africa (Walker 2006, and chapter 9 in this volume), or issues important to children (Biggeri, chapter 10 in this volume)—but in this chapter, I am exploring the universalist nature of Nussbaum's list, and applying it to Bangladeshi girls' perceptions of how their education is helping them to expand their opportunity sets and to reach their own valued achievements.

Capabilities: Flourishings and Failings

My study is linked to the work of a large Bangladeshi NGO: BRAC, established shortly after the country's liberation in the early 1970s, and well known internationally for its role in the education of girls and women, particularly in impoverished rural areas. My contact with the girls interviewed was through BRAC's nonformal Adolescent Development Programme, the overall goal of which is to improve the quality of life of vulnerable adolescents (BRAC 2004). My study included 17 adolescent girls, most of whom were attending secondary school when I first met them, and most were members of the Adolescent Development Programme, attending the *Kishori Kendro* (KK), or adolescent centers. The girls were from two poor rural districts in Bangladesh. In the first district, ten girls were interviewed in pairs or small groups, and in the second, seven girls were interviewed individually on two occasions six months apart. None of the girls was from an extremely poor family; less than 50 percent of children go on to secondary school in Bangladesh, and those are from relatively wealthy families. The interviews were loosely structured around their educational experiences (formal, nonformal, and informal) and plans for the future, and while no reference was made to capabilities or Nussbaum's list, my questions were shaped to elicit responses related to the development of the various capabilities.

What follows is a grouping of themes that emerged in conversation with the girls, roughly in line with evidence of capability flourishing or failings. Each theme highlights capabilities on Nussbaum's list, although there are overlaps between themes, and some slightly forced analytical compartmentalizing of capabilities. All names have been changed to protect the girls' privacy.

Learning Lives

Educational content and processes help shape the way we live our lives, and here I look at how the girls have learned to live the lives they lead, or will lead. Learning is interpreted in its very broadest sense, and includes what is learned outside any formal structure, within nonformal education such as that provided by BRAC, and formal and informal learning within the formal education system. Two Nussbaum capabilities that link most directly to education—*practical reason* and *senses, imagination, and thought*—are highlighted here. Nussbaum stresses the importance of being able to engage in critical reflection about planning one's life, and education should surely enhance this.

Almost all the girls had a strong belief in the power of education to bring about positive outcomes. Because the girls were still at school, there was no way of knowing if their education would help make their dreams come true, but what was evident was the fact that "being educated," or being seen to be educated was a source of enormous pride and confidence. These are girls from poor rural areas, many with illiterate parents. They are the first generation of girls going to school in significant numbers, and they are able to dream of a different, brighter future—as shown in this extract from an interview with two friends in Class 9 (age about 15–16):

> Janet: Do you think your education has made more positive possibilities for you, and if so, in what way?
> Khalida: Yes, it's had a great impact on my life. Now I can say what I want to do, and *if* I want to get married, and *when* I get married . . . but if I had had no education, I would just stay quiet, and let the others decide everything . . .
> Samira: . . . because we've been to school, we can talk to everyone and talk to strangers, and people who come from the cities, and we feel confident and we can speak up. We don't feel shy or anything.
> Khalida: Now we get to know about what's happening around, and we can hear or read about things when we go to the KK, and we can read the paper whenever we get one, and we can talk about these things. And like now, we can come to you, we can come to the office and talk to you people, but if we'd stayed home we would not be able to come out of the home and interact with anybody.

It is interesting to note that some of the things they value most—such as being able to determine when or whether to get married, or keeping themselves informed of current events—come from the KK (i.e., their nonformal education) rather than from their schooling. Issues such as dowry and child marriage are largely absent in the formal system, as is the space for such discussions. Again and again, asking girls what they valued most of what they had learned, the girls who attended KKs mentioned what they had learned there. Those who did not attend the KK were much less able to articulate what they had learned of value, and could not get much beyond naming their favorite subjects.

Despite the strong confidence in the role of education in improving their lives, there were many examples of the girls nonetheless learning their lower status as girls. One of the most obvious was in relation to subject choice. On entering Class 9, students are channeled into arts, commerce, or science streams. The science stream has the highest status, and is most likely to lead to good jobs. Statistics show a strong male bias in science subjects, and discussions with the girls indicate how this bias comes about. For example, in Class 8, Ayesha was ranked 16th out of 75 in her class, and when I first spoke to her she said she hoped to enter the science group. When I spoke to her six months later, she was doing arts. This was because a male cousin thought arts would be better for her. His arguments were financial: studying science generally means taking private tuition, which he thought her parents could not afford. So although Ayesha's examination results would have enabled her to go into the science stream, her parents took the advice of a young male relative (who was himself studying science) and decided not to invest in the more prestigious but more expensive science education. Ayesha accepted this decision, and seemed unaware of how this might affect her future. This was not the case with Roushan, who was in Class 10 when I met her, and was ranked 11th out of 70 in her class. Even though she knew she needed science for her career plans, she was pressured out of it by her teachers, her big brother, and her parents.

> Roushan: I wanted to do science, but there were no other girls who wanted to do science.
> Janet: Not one?
> Roushan: No, not one. There was only me . . . and the teacher asked me to do arts. Because there was no other female in the science group, only males doing science.
> Janet: So what about your guardians—did they want you to study science or are they happy that you're doing arts?
> Roushan: My brother advised me that if I took arts, it would be easy for me. He studies in science . . . He said that I did well in maths, but it's not only about maths. So it's better to choose arts, and the teachers also told me that it would be good for me.

Both her parents—the father with little schooling and the mother with none—deferred to the young brother on this. Roushan wanted to be a doctor, but knows this is no longer possible. She then decided she would like to join the police, but her mother is reluctant to allow her to do this because she has heard that candidates have to take off their clothes in order to have a medical, a thing that she cannot countenance her daughter doing. Roushan is well able to make reasoned decisions in planning her own life, but is blocked from carrying out her plans. She is learning that girls' choices are restricted. Her brother has learned that, too.

At least Roushan is able to make informed plans about her life, even if those plans are thwarted. Another girl—Shamima—following the arts course in her Secondary School Certificate examination year also had plans to be a doctor, and it was not until we talked that she realized that she needed science to get into medical college. It was obvious that she was not getting any sort of career guidance, and she was not able to use practical reason to plan her future. Once she learned that she was not taking the right subjects for a medical career, she decided that she would like to be a teacher. In fact, most of the girls I spoke to, when asked what sort of job they would like when they had finished their studies, said that they would like to be either a teacher or a doctor. Both are seen in Bangladesh as a natural extension of women's "caring" roles. The girls have been given a very limited set of options from which to make their choices.

Even though the girls talk of plans for working outside the home—and some of them do with great passion—they have learned that their main objective in life is to marry and have children. Girls' education is more a strengthening of the traditional roles of women than a preparation for paid employment. Girls do home economics (an option not available to boys) in order to learn better how to run a household, and how to look after their children. They acquire literacy in order to teach their children or to help them with their homework. Many of the girls also seem to have learned that their future husbands should have a higher level of education than themselves, although none was able to explain why this should be. These things are what the girls have learned about their future lives. Most of this learning has been part of the informal curriculum.

Taking or Losing Control

As shown in the previous section, even if girls are able to make plans for what they have reason to value, they are often not in a position to control those plans. This is partly because they are still minors, partly to do with limited financial resources, but perhaps mostly to do with the fact that they are girls. Agency is assumed in all of Nussbaum's capabilities; without agency there can only be basic well-being functionings (Sen 1987). The main focus here is on the capability of *control over one's environment,* which Nussbaum divides into two subcategories: *political* and *material*. As adolescents, the girls are not yet concerned with party politics; what is treated as "political" here is along the feminist lines of "the personal is political," and includes issues of great importance to adolescent girls in Bangladesh, such as child marriage and dowry practice. Material control is discussed in connection with physical or financial resources, and although Nussbaum talks of the right to seek employment on an equal basis with others, because the girls are adolescents, I am focusing on what is needed to get an education on an equal basis with others.

I have already shown that girls have little control over subject choices. They also have little control over the quality of their schooling, often being sent to lower-quality schools nearby while their brothers are free to go farther afield to enroll in a better school. Halima, although lucky in having a supportive older brother who persuaded their parents to let her take science, realizes that the school she is going to does not have high academic standards. However, she has no control over that. At first the reason appeared to be financial:

> Halima: The school's near my home, so I save money on rickshaw fares, and the good school in the town—if I went there I'd need more fees, and more money. But we're not so rich.

Further questioning revealed another reason:

> Janet: So which secondary school did your brother go to?
> Halima: [names a school in the nearby town] It's a good school.
> Janet: Is it boys only?
> Halima: No.
> Janet: But your parents were able to give him the rickshaw money?
> Halima: No, he had a bike.

Girls in Bangladesh are highly unlikely to have access to resources such as bicycles. This is linked to notions of "modesty," mobility, and the simple tendency to invest more in sons. But because Halima does not have the bike, she cannot choose to go to the school that she knows would be better for her. Several other girls I spoke to went to local madrassas (Islamic schools). Even though the academic standards of madrassas are generally lower, and many do not offer science, parents are happy to send their daughters to such schools because of concerns for reputation and security. Their sons are sent to better, mainstream schools farther afield.

For some families, meeting the cost of secondary education is a struggle. Most of the girls were receiving a small stipend while attending secondary school. The stipend is supposed to cover school fees, the cost of basic textbooks, and examination fees. However—because of factors such as the need to work or look after siblings, or the inability to pay for extra books or private tuition—those from the poorest families had difficulty in meeting the attendance (75%) and achievement (45%) criteria for receipt of the stipend, and their schooling was at risk. One girl had had to leave school because her family could not afford the costs, which rise sharply in Class 9 because of the need for more books:

> Akhtar: I want to continue with my school, but I have these financial problems, so I'm not able to.
> Janet: Right. Were you getting the stipend when you were at school?
> Akhtar: Yes, but it's not enough for my books and everything . . . so . . . it's not enough. We need a lot more money to continue with schooling.
> Janet: How much do you think you'd need for books and things for Class 9?
> Akhtar: Taka 2,500.

"Taka 2,500" is about GB £18 / US$ 16 (April 2007), a significant amount for a poor rural family. A girl is Class 9 is given Tk 250 toward the purchase of new books—that is, one-tenth of what is needed. Akhtar has two older brothers who were able to continue their schooling because they were able to do odd jobs to earn money. This covered the costs of their books and their private tuition. Akhtar was also earning a little money through rearing chickens and selling eggs, but she had no control over the money she earned; it went into the family pot, and she was pulled out of school. However, she is still allowed to attend the KK, where she is learning much of value to her, and which might help her have control over her environment in the future. There she has learned important things about the laws on dowry and child marriage, the importance of birth registration, and—most importantly—her rights. In addition, it has helped her build up a strong social network, and she has people outside the family that she can turn to if she has problems. That expanding circle of contacts is widening the possibility of control. As she says, despite her great disappointment about having to leave school:

> I know myself better. I understand things better, I can plan for myself, and I have hope for the future. I can hope for being on my own, being independent.

Making Bonds

That social network, and the confidence she feels in having that network, indicate that the capability of *affiliation* is strongly linked with educational processes, and with the development of agency. Nussbaum identifies two main types of affiliation: one is the development of friendships and empathy toward others, the other concerns being seen by others as a dignified being of equal worth to others. In this section I briefly look at how education is linked to affiliation. The capability for *emotions* can also be included here as they are strongly connected to human associations.

Nussbaum's second category of affiliation is very relevant in an educational setting, dealing as it does with self-respect and nonhumiliation, equality and nondiscrimination. I have already touched on how the fact of being seen to be educated increases the girls' sense of self-worth. Even if the education they are getting is not enabling them to have good careers, it has a feel-good factor, which enables them to interact with others on a more equal level. They can do little about the inequalities in and around their schooling, but like Akhtar they can draw strength from what they have learned from others, in or out of the family.

Affiliation with male relatives is a strongly recurring theme in many of the girls' stories. Sometimes that affiliation is negative (as with Roushan's cousin persuading her parents not to let her study science), sometimes positive (as with Halima's brother persuading her parents to allow her to). Halima's brother cannot be blamed for having a bicycle while she has not, and he has shown himself to be a strong support in her life in many ways. Both parents are illiterate, so they rely on him—the first educated person in the family—to guide them in their decisions. When I spoke to Halima, her brother had left their rural home and was studying at one of the best universities in the country. But he continued to write urging his parents to keep her in school, and he regularly sent money he earned giving private tuition to help defray the costs of her education. He had also coached her in the steps she needed to take to take up her chosen career (medicine). Another girl, Rasheda, also had a supportive older brother who sent money to pay her school costs, but unfortunately he died while she was in Class 6, and she was withdrawn from school. Rasheda is not a member of the KK, so when she was pulled out of school, her opportunities for affiliation became severely restricted.

Those who have had a longer period at school, and particularly those who attend the KKs, will have had more opportunity to forge bonds with people outside the family and to discuss things of importance to them, and having learned how to do this, will be better equipped to do so again later in life. One girl, expanding on what others had said, articulated the importance of the opportunities for affiliation, and the better opportunities at the KKs:

> Khalida: Yes . . . we [at the KK] have very long discussions. . . . But not in school. We go to school and we have all these friends, but we don't have

enough time to talk to friends. Because we have one class after another class. And there's only half an hour tiffin break. So what happens is, when we get to the KK, some of our friends from the school might be there, and some other girls are maybe from other schools, and there are some girls who don't go to school as well, who've dropped out. So all sorts of girls from the same village are gathering there—we have very lively discussions.

Once again, there is evidence of nonformal education offering the girls more of what they value, something that is not nurtured in the formal system. The KKs also offer (limited) opportunities for games. Affiliation is often strongly linked with play and sports, and while this may be so for boys in Bangladesh, this is not necessarily the case for girls, as can be seen in the next section.

Getting Physical

Three of Nussbaum's capabilities are directly concerned with the physical: *life*, *bodily health*, and *bodily integrity*. I am not addressing *life* here, although it is covered in KK topics such as the risks of child marriage, maternal mortality, or violence against women. The focus is on how lives can be limited simply by having been born female. I include the capability of play here, it being—in a society in which purdah is practised and mobility restricted—inextricably linked to bodily health and integrity. Connected to bodily integrity, which discussions indicate is a strong feature in girls' educational experiences, is the potential loss of affiliation on an equal status with others, to appear in public without shame, and a lack of appropriate resources, especially for adolescent girls.

Discussions with these adolescent girls highlighted one problem with resource-based approaches, and might partly account for girls' underachievement at school. This is simply to do with the fact that adolescent girls menstruate. Although all the girls I spoke to said there were separate toilet facilities for girls at school, for many of them these facilities were not adequate to meet their menstrual hygiene needs, which led in extreme cases to absences of up to one week a month. The policy of building separate latrines for girls does not make provision for those girls from poor families that cannot afford disposable sanitary wear. One very shy girl found it extremely difficult to explain her irregular attendance at school, but after much prompting she acknowledged it was linked to menstruation, and then spoke about it.

Janet: Have you ever been at school when your period started?
Hosne Ara: No, I never go to school when my period's due.

Janet: So you just wait at home when it's due?
Hosne Ara: Yes, I don't go to school. So, say, if the date is the 9th, and it's due on the 10th, I stay at home. So I've never experienced this at a school.

Not only was she unable to change her cloth at school, but she was also so frightened of having to deal with bleeding at school that she simply did not go. Such matters are not spoken of in the formal education system, but may be discussed informally if there is a woman teacher. In schools where there is no woman teacher to talk to, girls may be forced to lie. One girl said that if she needed to go home to change her cloth and there was no woman to get permission from, she would tell the male teacher she had a headache.

Bodily integrity was a recurring theme in the discussions with the girls. This has been touched on earlier in relation to concerns for the girls' security; and sexual harassment is widespread. There are frequent reports in the newspapers of girls being harassed, abducted, raped, or even killed on their way to or from school. For one girl, this fear led to her withdrawal from school.

Farhana: I was in Class 9, I completed it. I haven't been in school for two months, but I hope to continue with it.
Janet: And why did you leave school?
Farhana: I used to go to school with a friend of mine. She had some problems and she delayed it, so I've been away for two months. I have to wait for her, and we're going to go to school together again.

She was pulled out of school simply for lack of a friend to walk to school with. Although she hopes to go back to school, having lost two months of Class 10—the big public examination year—her chances of success are significantly reduced.

Other aspects of bodily integrity have been referred to earlier in relation to the raising of hands, and the lowering of eyes. This extract is from a discussion with two girls on how they deal with harassment from boys:

Khalida: And what we do is we just try to avoid them and look at the ground while they're walking and not make any eye-contact with them. Just avoid them.
Janet: Did somebody teach you? Did somebody tell you you have to keep your eyes down?
Samira: No, no one told us.

This behavior has been part of their informal education. They have learned that boys can be troublesome, and they have learned strategies for dealing with this. But of course, while they are looking at the ground, they are missing out on so much else that is happening in the world. Girls also miss out on play, which seems to stop at puberty. They learn shame, and embarrassment about their developing bodies—also part of their informal education. The more I spoke to girls, the more I realized how tragic it was that this source of pleasure—available to boys and men—was cut out of their lives so early. Sports and games are not part of the compulsory curriculum in Bangladesh, but most schools offer "co-curricular" activities—with gendered differences:

> Farhana: We don't usually have games like the boys. Girls do singing. And we have a cultural event once a year.
> Janet: Would you like to do the games and sports like the boys?
> Farhana: The teachers don't allow it. We'd like to play, but it's not done.
> Janet: Why isn't it done?
> Farhana: We can't. Because boys would gather round when we play, say, if we have some volleyball, and they'd—you know—bother us.

Boys also have the chance to play games such as cricket or football during the breaks at school, while the girls stay inside. The KKs offer more opportunity for play, but this is mostly restricted to sedate indoor games such as Ludo or chess. As Farhana goes on to say, people from the village would not like the idea of "big girls" running around in the open. The physical fact of being born a girl has a great impact on the richness of the educational experiences available. It affects the way they learn to behave in class, and how much attention they get from the teacher. It determines how they spend their break times and how—and how far—they travel to and from school. It also increases the likelihood of absence from school. All this adds up to a severe limitation of a range of capabilities.

Summing Up

The picture painted here may seem very negative, but my intentions are not. Bangladesh has made remarkable progress in getting girls into schools, and just being in school gives an enormous boost to girls' self-esteem. But getting girls there is just the beginning. We have to consider what it is that girls are learning there—including and beyond the national curriculum—and how their capabilities are being developed or denied. As Nussbaum says, "Years of schooling, everyone would admit, are an imperfect proxy for education" (2002, p. 73).

While in this analysis there have been problems with the indivisibility of capabilities (e.g., the gendered aspects of being a girl arising in all areas), and some of the capabilities on Nussbaum's list have been neglected or omitted, some patterns do emerge. There seems to be strong evidence of schooling/education enhancing the capability of affiliation. With this, as with many other capabilities, the evidence is stronger in the nonformal education provided by BRAC. Other areas are more ambiguous, and in many cases can be seen as capability failures.

There are strong indications that the two Nussbaum capabilities most clearly linked to education are not being developed fully, as evidenced, for example, by the girl who was not able to use *practical reason* to plan a career as a doctor because she didn't know she had to study science, or by the fact that girls can do little to plan their own lives as so much is beyond their control. With senses, imagination, and thought, which Nussbaum stresses should be by no means limited to literacy and basic mathematical and scientific training, we have evidence of girls being excluded from the sciences, and little evidence of formal education beyond the basics. However, the KKs do encourage extensive reading, discussions, music, and dance. The capability of life is presumably enhanced through education, but again more strongly through the nonformal program that more adequately covers the issues girls need to learn about.

Bodily health is also more comprehensively covered in the nonformal system. Physical education for girls is largely sidestepped in both formal and nonformal systems for girls, while informally the girls learn that their bodies are sources of shame and embarrassment, which compromises their bodily integrity. The important issues of sexual harassment and violence are also not part of the formal curriculum. Linked to this—especially for girls—is the capability of control over one's environment. As minors, and as girls, they have little control, but the KKs do at least cover the potential for control through teaching girls the laws and their rights. The most obvious capability failure is that of *play,* most noticeable as girls reach puberty. This in turn leads to a reduction in bodily health, and more importantly, in affiliation opportunities.

If one sees education as a preparation for adult life, and if we agree that such capabilities are needed for truly human functioning, then an education system should be framed to help develop those capabilities. In Bangladesh, the formal system as it stands is as much about capability deprivation as development, and there is an urgent need for educational reform. Building on what has been developed by NGOs such as BRAC would be a good place to start, and Nussbaum's list could be very useful for identifying those capabilities enhanced by educational processes, those where education enhances conversion abilities, and those capabilities that are excluded from the education system. That is, in identifying the role of education in the flourishing or failure of capabilities, and where necessary, modifying the curriculum and approach to reduce the failures and nourish the flourishing. In short, education cannot be regarded as a basic capability unless it specifically addresses the process of developing the capabilities necessary to live a life one has good reason to value.

References

Ahammed, Kazi Farid. 2003.*Country report on accelerating progress on girls' education in Bangladesh.* Bangladesh Ministry of Primary and Mass Education. Available from http://www.unesco.org/education/efa/global_co/working_group/WGEFA4_Bangladesh.ppt#1 (accessed April 5, 2005).

Arabic News. 2005. *UN General Assembly president urges member states to deliver on summit declaration.* Arabic News Available from http://www.arabicnews.com/ansub/Daily/Day/050917/2005091724.html (accessed September 19, 2005).

Arends-Kuenning, Mary, and Sajeda Amin. 2001. Women's capabilities and the right to education in Bangladesh. *International Journal of Politics, Culture and Society* 15 (1): 125–142.

Asia-Pacific Daily News Review. 2005. *Sri Lanka reconfirms her participation in Dhaka SAARC Summit.* Office of the UN High Commissioner for Human Rights, Regional Office for Asia-Pacific Available from http://www.un.or.th/ohchr/SR/News/2005/DailyNewsBulletin/9/19%20September%202005.doc (accessed September 20, 2005).

BANBEIS. 2005. *Educational information and statistics* [November]. Bangladesh Bureau of Educational Information and Statistics. Available from http://www.banbeis.gov.bd/index.html (accessed February 2, 2006).

———. 2006. *National Education Survey (Post-Primary)—2005, final report.* Dhaka: Bangladesh Bureau of Educational Information and Statistics. Available from http://www.banbeis.gov.bd/report/report.pdf (accessed February 5, 2007).

BRAC. 2004. *Adolescent journeys: Building the roads.* Dhaka: BRAC (Adolescent Development Programme).

Center for Women's Global Leadership. 2005. *United Nations 2005 World Summit outcomes: Gains on gender equality, mixed results on poverty, peace, and human rights* Beijing and Beyond. Available from http://www.beijingandbeyond.org/documents/FinalSummitReportBack2.htm (accessed Janaury 15, 2006).

DPE. 2002. *Primary education statistics in Bangladesh—2001.* Dhaka: Primary and Mass Education Division, Government of Bangladesh.

Jeffery, P., and R. Jeffery. 1998. Silver bullet or passing fancy? Girl's schooling and population policy. In *Feminist visions of development: Gender analysis and policy,* edited by C. Jackson and R. Pearson. London: Routledge.

New Nation. 2005. 15,631 students get GPA-5 in nine education boards: 54 pc pass SSC exams. July 9, 2005.

News from Bangladesh. 2005. *Bangladesh finds World Summit fruitful: Reaz.* News from Bangladesh Available from http://www.bangladesh-web.com/news/view.php?hidDate=2005-09-20&hidType=NAT&hidRecord=0000000000000000061562 (accessed September 20, 2005).

Nussbaum, Martha C. 2000. *Women and human development: The capabilities approach.* Cambridge: Cambridge University Press.

———. 2002. Women's capabilities and social justice. In *Gender, justice, developments and rights,* edited by E. Molyneux and S. Razavi. Oxford: OUP/Oxford Studies in Democratisation.

———. 2003. Capabilities as fundamental entitlements: Sen and social justice. *Feminist Economics* 9 (2–3): 33–59.

Raynor, Janet, and Shamima Pervin. 2005. *Observing lessons through a gender lens.* Dhaka: PROMOTE.

Robeyns, Ingrid. 2003. *Sen's capability approach and gender inequality: Selecting relevant capabilities.* Available from http://www.st-edmunds.cam.ac.uk/vhi/nussbaum/papers/robeyns.pdf (accessed April 1, 2005).

———. 2006. Three models of education: Rights, capabilities and human capital. *Theory and Research in Education* 4 (1): 69–84.

Sen, Amartya. 1987. The standard of living. In *The standard of living,* edited by G. Hawthorn. Cambridge: Cambridge University Press.

United Nations. 2000. *UN Millennium Development Goals (MDG).* Available from http://www.un.org/millenniumgoals/ (accessed July 18, 2004).

———. 2005. *Resolution adopted by the General Assembly: 2005 World Summit outcome.* New York: United Nations. Available from http://daccessdds.un.org/doc/UNDOC/LTD/N05/511/30/PDF/N0551130.pdf?OpenElement.

UNDP. 2005. Human development report 2005: International cooperation at a crossroads—Aid, trade and security in an unequal world. New York: UNDP. Available from http://hdr.undp.org/reports/global/ 2005/pdf/HDR05_complete.pdf (accessed July 7, 2005).

Unterhalter, Elaine. 1999. The schooling of South African girls. In *Gender, Education and Development: Beyond access to empowerment,* edited by C. Heward and S. Bunwaree. London: Zed.

———. 2003. The capabilities approach and gendered education: An examination of South African complexities. *Theory and Research in Education* 1 (1).

Walker, Melanie. 2006. Towards a capability-based theory of social justice for education policy-making. *Journal of Education Policy* 21 (2): 163–185.

Zia, Begum Khaleda. 2005. High level plenary meeting of the UN: Statement by Her Excellency Begum Khaleda Zia, Prime Minister, People's Republic of Bangladesh. Paper read at the 2005 World Summit, September, at New York. Available from http://www.un.org/webcast/summit2005/statements/ban050914eng2.pdf (accessed January 21, 2006).

9

Selecting Capabilities for Gender Equality in Education

Melanie Walker

The chapter considers how to apply the capability approach in education by selecting a list of education capabilities with a particular focus on gender equality. I combine a top-down theoretical approach using ideas from the capability approach, with a grounded view from education policy and from what a group of South African girls say matters in their lives. I then extrapolate from this a list of valued education capabilities. The argument here is that if we are to address the specificity of gender equality in education, we need a way of evaluating when we have gender equality and quality in educational provision, processes, and outcomes. Learning is fundamental in education, and the capability approach resonates with approaches to the significance of learning in shaping our "powers and capacities, in our unfolding agency" (Ranson 1998, p. 19). Developing learners' capabilities and agency further resonates with research that points to the importance in contemporary times of being able to negotiate our lives reflexively, to be able to change and adapt, and to make good choices in navigating our lives (Wynn 2004). Learning is then "the transformation of understanding, identity and agency ... the capacity to develop and sustain reflexivity" (Edwards, Ranson, and Strain 2002, p. 533).

There can be no doubt that formal education provision matters. To be educated is a capability in itself while education is also made up of a number of separate but intersecting and overlapping constitutive capabilities (e.g., knowledge, literacy, social relations). Education capability also impacts on the formation of other valued capabilities in girls' lives (see Terzi, chapter 2). This is not to say that informal learning opportunities are unimportant in contributing to our educational development. But formal education, particularly at the level of compulsory schooling, is a crucial site for reproducing and transforming social norms and culture and for identity formation (who we take ourselves to be), which identities and abilities count (and which are devalued), and what we see as possible for ourselves. Furthermore, access to credentials gained through formal education shapes our economic opportunities and choices. Generally having more formal education is then an advantage. A good quality education is of intrinsic worth in our personal development, and instrumental in opening up economic opportunities. Education is constitutive of other aspects of human well-being, such as having confidence in one's abilities as a life long learner, or good health, or civic participation. Having education, therefore, potentially enables other capabilities.

Because education is so central in the distribution of personal, social, economic, and political goods in society, equality in education matters (Baker et al. 2004). It further matters that experiences of formal education should be positive, enriching, and life enhancing for all learners. In education we also need to keep in mind Sen's (1999) argument that emphasizes not only the freedom a child may have in the present, but also the freedom the child will have in the future. If a child refuses, or is denied or restricted in her access to the goods of education, this will in return reduce opportunities in adult life and restrict her future individual freedom and agency. As Baker et al. (2004, p. 143) emphasize, for children in school, their education is experienced not so much as a preparation for life but "it is life itself." Lifelong education then begins with the young girl. Learning is seldom linear and immediate; it is more often recursive. New learning builds on existing knowledge, past experience, and pupil identities. Learning is then a process of "becoming" as well as "being" over a life course, and through cycles of schooling. It is, therefore, a matter of concern that schools can unfortunately be places both of freedom and unfreedom (Unterhalter 2003). Capabilities can be diminished as well as enhanced so that poor quality schooling is a disadvantage, and one which might persist throughout a lifetime on people's perceptions of themselves as learners (Gallacher et al. 2002). Sen (1992, p. 44) is, therefore, right to argue that (equality) education is one of "a relatively small number of centrally important beings and doings that are crucial to well-being."

Selecting Capabilities for Gender Equality Assessment

I now apply the capability approach to produce a provisional, situated list of education capabilities, with specific attention to achieving gender equality in South African schools. I follow Robeyns's (2003) criteria for the selection of capabilities by making my selection explicit and public, and my method clear, by first producing an ideal list, and by listing capabilities that are not reducible to each other. The list is produced in five steps. The first is to briefly consider the capability approach and the specificity of education. I then consider post–1994 education policy, in my case of South African schooling, and in particular, matters of gender equality. I then ground my list in the experiences of South African girls, following by engaging with other lists of capabilities, searching out those elements that might apply specifically to education to see what overlap exists. The final stage involves debating the list with others by publishing my list here.

Importantly, mine is an ideal-theoretical list, illustrative in intent, and cannot substitute for participation and dialogue in a deliberative and democratic process of policy making. The issue of lists is contested (see chapter 1). Nussbaum (2000) has produced a list of ten central universal human capabilities. But Sen himself argues for the importance of democratic participation in deciding on valued capabilities for public and education policy because public communication and participation is both valuable in itself and valuable for the subsequent scrutiny of policy implementation, its success and failures (Sen 2004). I briefly want to make two points. First, that I have here a list, not one central universal list for all contexts. A working list such as this offers some purchase on what is to count as education for gender equality. Unless we do this we risk saying, as Nussbaum (2003) argues, that we are for equality and justice (in education) but that any old conception of equality and justice is all right with us. Moreover, in combining gender equality capabilities with education capabilities in a multidimensional list in which all the capabilities matter, we demonstrate that gender equality and education equality are interwoven, one within the other. We cannot then have a quality education that is also not an equality education. From the perspective of the core educational practices of curriculum, pedagogy, and assessment we must then ask who has the opportunity to develop valued education capabilities, and who has not?

Second, this list involves selection. The issue here is that to address the specificity of education in the human development paradigm, and to argue for the development and distribution of capabilities for and as equality in education, we need to select what we take to be education capabilities, such that if some or all of these were not present we might not have a process of education (and see Walker 2006). We can also argue that because the capability approach attaches great importance to agency and to genuine reflective choice, that education in any context should promote agency, and as a key element of this agency that education should facilitate the development of autonomy and empowerment. Sen ascribes an instrumental role to education for realizing economic opportunities that we might provisionally describe as a capability for paid work (in formal and informal economic spheres). He attaches importance to the relations a person has with others, to the social role of education and generally the capability to be a full participant in society; so a capability for social relations is arguably important. In the context of schooling such social relations might take the form of an institutional culture, supportive teachers, and care from and to peers in learning arrangements. Thus we already have three provisional education capabilities: *personal autonomy, paid work,* and *social relations.*

Selecting Capabilities from Education Policy

I turn now to the specific context of South Africa in order to extrapolate capabilities from education policy. The post–apartheid 1996 Constitution guarantees to everyone the right to a basic education; schooling is now compulsory for all children for nine years. The National Policy Education Act envisages an education system that contributes to "the full personal development of each student", and to citizenship for the building of a democratic nation. It includes "the promotion of gender equality" and "encouraging independent and critical thought" (South Africa 1996, p. 4). The act further identifies a purpose of education as "the advancement of knowledge," and encourages "capacities necessary for reconstruction and development and knowledge and values for citizenship." It additionally introduced a language policy to replace the two official languages of apartheid South Africa (English and Afrikaans) with the formal recognition of all 11 languages spoken in South Africa. This translates into a commitment to the right of every child to be taught in the language of his or her choice, where this is "reasonably practical" (Department of Education 2003), and to a vision of education contributing to cultural recognition development. Although not without its critics (e.g., Meerkotter 1998), curriculum revision for schools culminated in Curriculum 2005, which is underpinned by principles of learner agency ("active learners," "learners take responsibility for their own learning," "learner-centeredness"), a thin autonomy ("critical thinking, reasoning, reflection and action"), respect ("constant affirmation of their worth"), and development ("what the learner becomes and understands") (see Meerkotter 1998, p. 59).

In 1996 a Gender Equity Task Team was also established to make specific policy recommendations on achieving gender equity (Wolpe, Quinlan, and Martinez 1997), with the department of education committing itself to a gender-sensitive education system for the development of a nonsexist South African society. The Tirisano education plan, adopted in 2000, has as stated policy, "To promote values, which inculcate respect for girls and women and recognize the right of girls and women to free choice in sexual relations" (DOE 2000, quoted in Unterhalter 2003, p. 16). Additionally, corporal punishment for boys and girls, once widespread in black schools, has been officially banned. Such measures are particularly important given the widespread harassment of girls at school, both by fellow students and by teachers (Unterhalter 2003).

The following important capabilities can be extrapolated from education policy in South Africa as constitutive conditions of the basic capability to be educated:

1. *Independent and critical thought,* critical thinking, reasoning, reflection, learner agency and responsibility for their own learning (a thin personal autonomy)
2. *Knowledge* for values, citizenship, contribution to economic development
3. *Bodily integrity* and health, safety at school, no corporal punishment, freedom from sexual harassment and violence, choice in sexual relationships, protection against HIV
4. *Respect* for self, for others, for other cultures, being treated with dignity (a form of social relations)

Selecting Capabilities from Girls' Lives and Voices

Turning now specifically to girls schooling, based on statistics of access and school-leaving (matriculation) examination outcome, in South Africa they are doing reasonably well. According to the Gender Parity Index in 2001 there were 5 percent more boys at primary school, and 10 percent more girls in secondary schools. In 2001 more girls wrote the school-leaving matriculation examination, but the relative pass rate of boys was slightly higher. The national matriculation rates in 2001 showed a pass rate of 60.1 percent by girls and 63.6 percent by boys. Boys are doing better overall and better, even if by a small margin, in certain subjects (DOE 2003). If these statistics were taken as the informational space to address "equality of what," there would appear to be no significant gender inequality. But the information does not tell us about the factors of class, gender, and race in student achievement, nor about the experiences of girls and boys in and out of schools, nor about the success rates of different kinds of schools. They provide important but only very preliminary information to make judgments about education capabilities.

Thus the next section of this chapter draws on interviews conducted with 40 girls aged between 15 and 16 years in four Cape Town schools in August 2003 to establish what capabilities these girls value to live the lives they choose.[1] Of the 40 girls, 5 were white, 16 African, and 19 of mixed race, reflecting the demographic makeup of Cape Town and the Western Cape, with its large mixed race population. They attended four schools[2] with differing racial and social class intakes, and differing levels of resources supplementing the same basic state provision. The girls' socio economic backgrounds were also diverse. Some were the daughters of professionals, some had parents who were unemployed or working in unskilled occupations. Their homes varied from small shacks in squatter communities, to brick-built two-room township homes with outside bathrooms, to solid, comfortable homes in middle-class suburbs. Some had access to a computer at home and a room or a quiet place to study. Others shared rooms with relatives and had little or no private space. The 40 girls valued independence and saw their schooling as contributing to this, saying things such as, "I want to be independent ... do your own thing, like for freedom" (Zurina). They valued many different dimensions of well-being and held diverse views of what for them was the good life, shaped socially and by individual circumstances. For example, they varied in the importance they attached to religion and spiritual practice, to the careers they hoped to pursue, to their valuing of material wealth, their desire to engage in public service, their desire to marry or not, their involvement in aesthetic activities, their love of nature and animals, their interest in fashion, and their desire to have "fun." This diversity of views then supports the capability of autonomy, which facilitates reflective choices for diverse forms of a "good life."

Education plays a key role in gaining independence and also in opening up economic and career opportunities—"I have to learn so that I can work and support myself," said Lumka, while Yasmina said: "The main thing of going to school is so that you can learn and go study [further] one day." For the African working-class girls at School A, in particular, schooling offered independence from male partners, and the choice not to marry. Tozi, for example, said that what was important to her was that "we are free to do what we really want, to be something you want to be because there is not anyone who is going to force you ... in these days if I don't want to be married, I will stop and say I don't want to, so I have the right to say I don't want to do this, I'm going to do this, according to my own needs and wants." A number of the African girls said that as educated young women with opportunities through their education to participate in the labor market, they were not willing to do all the domestic labor for a husband "who has two arms, two legs, two eyes, a mouth and everything" (Nombulelo). In speaking about their futures, all the girls offered aspirational narratives about their hopes. In particular, the girls at School A expect their futures to be significantly different from those of their mothers' generation.

Most of their mothers had been or were domestic workers with minimal education, and their fathers unskilled or semiskilled workers. Thus Thandi commented: "In the old days, if you were a girl, then boys they can make you their wife and you will have no more education. So now we are living a new life, we feel no one will tell us what to do, except yourself. In the old days our mothers were forced not to go to school, but to be a wife." Tozi then added: "I'd like to say in these days there are opportunities to do what you really like because in the past our mothers were domestic workers, there was no other kind of work they were doing. They couldn't go further at school because of money, they were poor. But in this case we have opportunities to go further; our parents try to educate us."

What emerged from these future plans was the importance of the capability to aspire for all the girls, which might be seen as commensurable with autonomy and planning a life, but which might also arguably stand as an important education capability in itself, not entirely reducible to any of the others. For example, one could argue that the capability of autonomy is developed through aspirations and aspiration is then reducible to autonomy. But one might also argue that the capability for paid work is fostered through aspirations for a career. I am, therefore, arguing to list aspiration as a separate capability because of its critical importance in redressing women's and girls' adapted preferences that leads them to settle for second or third best, or not to aspire at all. Appadurai (2004) argues strongly for the need to strengthen the capability to aspire. He suggests that this capacity constitutes a resource for poor people (and women, I would add) to contest and alter the conditions of their own welfare. Thus Appadurai argues for an expansion of people's aspirational maps, what we might call a thick rather than a thin capacity. I would further wish to incorporate the capacity for hope, which all these girls demonstrate, into the capability for aspiration. As Hage (2001) argues: "Hope is not related to an income level. It is about the sense of possibility that life can offer. Its enemy is a sense of entrapment not a sense of poverty" (p. 3). Education as a basic capability potentially fosters this important capability of aspiration. For girls school ought to make available new aspirational possibilities, and expand their horizons for action and life. What was most striking was the way all the working-class girls imagined better lives for themselves than had been possible for their parents. Their positive aspirations say: "I can be this person and do this sort of job," and so fracture cycles of adapted preferences in which girls and women settle for less. Aspiration produces new possibilities. Without the opportunities to go to school, and to complete 12 years of school, girls would find it difficult to imagine alternative futures to that of their parents, and harder still to realize those futures in their lives. In a country that for so long denied and diminished the aspirations of the majority of its population, and a country in which femaleness is still less valued than maleness, this is hugely important.

Appadurai links the capacity to aspire to voice, arguing that each reinforces and nurtures the other. Voice also supports the capability of autonomy. The intention again of listing it separately underlines the specific importance of the capability of voice for girls and women. Indeed, one might argue that women who have the mature capability of autonomy may well still struggle with their assertive voices, as studies of academic women suggest (see for example Walker 1997). Thus we begin to see how the capabilities are interdependent yet not strictly commensurable, and how the absence of one diminishes another. Where schooling fosters voice, here understood as the capacity to debate, contest, inquire, and participate critically, it simultaneously nurtures aspiration. Where children might be denied a capability for voice at home or in society, or where their aspirations might be cramped outside of school, there is then a particular ethical responsibility for the school to challenge exclusion, not to perpetuate it. This is especially important for girls from poor homes, but in a sexist and paternalistic society such as South Africa, it is arguably important for all girls. It therefore seems that the capability of voice is fundamental to education. Sometimes this can be quite literally the ability to speak out and speak back, as Nombulelo points out, "If you know how to talk for yourself, you can change things for yourself". She then added that black people had struggled "for freedom, for their own voice to be heard by everybody." Voice and agency intersect with and support each other. We also need to bear in mind that there are different voices available to us, and the female voices, as feminists have pointed out, that are accepted and recognized are voices that are pleasing, obedient, docile, supportive, and submissive, rather than agentic and assertive. Under conditions of masculine power in general, and the specific conditions in South Africa of harassment and violence toward girls and women and of HIV/Aids, girls also need the capability to be assertive, and powerful voices to speak back, to challenge, and confront. We see some of this emerging from the girls I interviewed, who are learning to stand up for themselves in school, to speak back to boys who show a lack of respect.

The capabilities of voice and aspiration are something on which other capabilities can all build. The capability of voice, one's own and hearing that of others, is also one to be exercised in class through the curriculum and pedagogy, challenging the continuation of practices of silencing and passivity. Voice might also find expression in confident participation in learning and in dispositions to learn which are strengthened by successful participation such that knowledge is then gained. As Appadurai suggests, we need to provide opportunities to practise these capabilities, that is, opportunities to function. By contrast pedagogies of silencing and passive learning do not contribute to voice, aspiration, or autonomy.

The capacity for friendship emerges from these girls' experiences as being much valued, including the differences among themselves: "We're so diverse and it's wonderful" (Helen); "We're different but we kind of relate to one another" (Nadia). Others value what they have in common as well: "I like the fact that we're so much alike. We can delight in each other in so many ways and I can speak to her openly" (Janine). Sharing and the quality of being "warm-hearted" are seen to be important values in friendships: "She's doing everything for me when I need help and I do everything for her when she needs help" (Kholiswa). Or bringing out different sides of somebody: "I'm very shy and they are the opposite of me and I like that because they bring out that part of me, when I'm around them I also talk and become lively" (Lillian). And offering good advice: "Every time I am in trouble she seems always to have the answers and every time she is in trouble, I have the answers" (Lillian). Loyalty and respect are important: "She's a very loyal person, and she's good at keeping secrets, she always listens to me" (Miriam). The idea of "having the qualities of a friend" is valued in others and in themselves. School provides significant opportunities to develop the capability and functioning for these kinds of social relations. However, there is another side to this capability for friendship. Except for School A at which all the pupils were African, there was limited association across ethnic and racial lines at the other three schools. Mostly this did not seem to operate as a deliberate exclusion, but girls did not necessarily go out of their way to form friendships with those from different racial groups. Often this was a matter of language—the African students would speak amongst themselves in Xhosa and without knowledge of the language, communication outside the classrooms was difficult. Close friendships could also contribute negatively to the development of positive learning dispositions where girls sought out others like themselves who had no interest in their schoolwork, thus in turn reinforcing their mutual disaffection.

School provides girls with access to subject knowledge, which will enable them to make future career choices, or simply enjoy this knowledge as an intrinsic good. In the South African school system, pupils choose their six matriculation subjects at the end of year 8. Here is just one example of the curriculum and knowledge opportunities and choices that school opens up for these girls. Pauline explains her subject choice of biology: "Because I really enjoy it, I like learning about the plants and the human body and things like that and geography because I found it easier, it was easy to catch on, and science because I wanted to be a doctor. But not anymore but it's still a good subject to have, it opens doors for you and maths because maths is a really helpful subject." The knowledge gained at school may be intrinsically valued, instrumentally valued (work), or positionally valued (a better university, expanded career options). Having this knowledge and the credentials that would not be possible without it, expands opportunities, agency, and freedom. But there is another side to school knowledge—the difficulty of knowledge-diminishing capability, where girls construct deficit identities for themselves when they are not succeeding in a key subject, or doing as well as their friends. Thus Megan at School C commented that all her friends were finding mathematics "easy ... then I don't understand why I can find it so difficult, so I just thought I was dumb." There is also the problem where pupils may have to study a subject they resent deeply, or when a subject they would very much like to study is not available. In the particular instance cited here, at School B, two of the African girls complained about having to study Afrikaans.[3] Additionally, they resented the fact that their own home language, Xhosa, was not even offered as a subject at the school.

Moreover, school lessons can undermine learning as well as support it. There is the boredom with the "same teachers who just drown you in their words; it's horrible" (Joanne), which might lead children to exercise their agency in ways that counter their own best interests. Teachers who leave classes to collapse into chaos so that little effective learning takes place do not enhance capability: "I mean to be honest with you, my history class ... I don't think she knows how to teach ... we're already in that phase where we really don't care anymore because it's so late in the year" (Pearl). Pearl and Sandra pointed out the gender dynamics of their classrooms in which boys end up getting more attention from teachers and from the girls as well: "The girls behave better in class. The boys generally do get more attention because they make a noise so they get attention from us as well. Most of the time we scream at them to keep quiet" (Sandra). The point is that achievement, aspiration, and voice can be compromised by schools. Two of the African girls at School B felt the (white) teachers sometimes expected too little of them just "because we are black" (Lillian). The school and its teachers for these two girls are not fostering their educational achievements as fully as they should, or as they do for other girls. Voice is also reduced by teachers who take unfair decisions about girls' ability, for example, to move girls from a higher to a lower grade class without consulting them. Sandra recounted how she had not done well in recent half-yearly mathematics examination so her teacher downgraded her from the higher grade to standard grade (an easier examination but not recognized for university entrance). As Sandra said: "The teachers don't say this is the reason why, they just tell you, 'you must go.'"

For all of these girls, their capability set is compromised by the sexual harassment, predominantly outside school, but for some inside school as well. While none of the four schools was characterized by the violence in many South African schools, three of them were nonetheless in some degree marked by a low-intensity, pervasive harassment of inappropriate touching and disrespectful behavior. None of the girls I spoke to found this acceptable. As Sibongile said: "A boy must talk to a girl nicely and treat her as a human being." At the all-girls school, the girls welcomed the opportunity to learn away from boys, said Lillian, "now that we are growing up." African girls who had friends or relatives at township schools also mentioned the continued use of corporal punishment at these schools as a reason for not wishing to go there. Indeed, the girls at School A remarked on how different their school was from other similar schools. There was no sexual violence or gangsterism at the school; they felt safe while at school, surrounded by a high fence, a locked gate, and a guard monitoring who comes and goes. As one of the girl said when asked why she did not attend a high school near to her home: "It's not a good [safe] school for me; there was a boy who was shot by the gangsters so I said when I go to high school I cannot go there because I don't think it's safe. So my father's sister said that I should come to this school 'cos it's a quiet school" (Kholiswa). It was clear that these girls valued being safe at school; it made a difference to their learning.

What then emerges is that these girls valued *learning* and the opportunities education will open up for them. They valued the *knowledge* they gain both from the subjects they enjoy and those they find harder or less interesting, and how their schooling qualifications would open up valued *economic opportunities*. They valued being *respected;* having their voices and languages acknowledged and recognized; and find teacher support, care, and appropriate expectations affirming and confidence building. They valued their *friendships* in school and reject their *bodily integrity* being encroached.

Engaging with Capability Lists

My next step is to consider other capability lists to see where they might overlap with my ideal-theoretical education list, even though these other lists encompass the whole of human development and not just education. I focus briefly on Nussbaum (2000), Narayan and Petesch (2002), Robeyns (2003), and Alkire (2002). Nussbaum (2000) has produced a list of ten core capabilities (see pp. 78–80); a threshold level of all the capabilities taken together is essential, she argues, to a life worthy of the dignity of the human being. Her list claims to be universal and cross-cultural. Failure of capability in any one aspect would be failure to live a fully human life. Setting aside the importance for Nussbaum of each and every component on her list (nothing can be left out), we might still consider which of the capabilities might be described as education capabilities, while still acknowledging the importance of all those on the list, should we so wish. Those capabilities she identifies that overlap with my list include capabilities of "practical reason" (being able to plan one's life), "affiliation" ("being able to live with and toward others, to recognize and show concern for other human beings, to engage in various forms of social interactions; to be able to imagine the situation of another and to have compassion for that situation; to have the capability for both justice and friendship.... having the social bases of self-respect and non-humiliation; being able to be treated as a dignified being whose worth is equal to that of others"); "senses, imagination and thought," "emotions," "bodily health," and "bodily integrity."

Narayan and Petesch's (2000) research with 60,000 interviewees also presents a list of ten capabilities, many of which overlap with Nussbaum's list and eight of which seem relevant to educational processes: "bodily health," "bodily integrity," "emotional integrity" (freedom from fear and anxiety, love), "respect and dignity" (self-respect, self-confidence, dignity), "social belonging," "cultural identity," "imagination," and "information and education." In her evaluation of a literacy project in Pakistan, Alkire (2002) generates a list of capability impacts in which at least four (empowerment, knowledge, work, relationships) seem relevant to educational processes. The capability of empowerment identified by the women as of value has some features in common with the notion of autonomy. Thus the women identified the importance of being able to solve their own problems, of deciding for themselves what is good or bad. Robeyns (2003), in her proposed list of 14 capabilities for gender inequality assessment, mentions a number of capabilities relevant to education, namely "education and knowledge" (having the freedom to be educated and to use and produce knowledge), "respect" (enjoying the freedom to be respected and treated with dignity), "social relations" (being able to be part of social networks), and "bodily integrity and safety." She is careful to point out that the capability of education should focus on more than just credentials and degrees; it should also pay attention to processes in schools and classrooms that produce gender inequalities.

Might there be capabilities drawn from those listed by these writers that could be added to my list? Nussbaum's (2000, p. 79) "emotions" (which includes "not having one's emotional development blighted by overwhelming fear and anxiety") adds detail to what we take emotional integrity to involve in schools and how it intersects with respect. How we feel affects how we learn, or fail to learn. For example, the girl who is told she is "too thick" to learn history, or comes to believe she is "too stupid" to grasp poetry, or says, "I'm not clever, I don't understand anything," or is so fearful of being beaten that she wants only to flee the classroom. Or the girl so upset at being publicly humiliated for some mistake or failure of understanding that no learning can follow. Emotions also have a positive impact on learning where they are a key source of understanding and awareness, as Baker et al. (2004) suggest. "Intellect," they say, "without emotions is harsh, because it is indifferent to the feelings and needs of others" (Baker et al. 2004, p. 165). Through developing our emotional capabilities, we develop empathy, caring, and solidarity. We might also wish to place greater emphasis on imagination as an element allied to emotions and emotional integrity. While autonomy develops our capability to own and take responsibility for our own lives and the consequences for and on others, emotions, respect, and recognition mean developing our capability for showing consideration to others, for understanding them, to participate in the human condition. If we take education to be a process of becoming and being a full human being, evidence of the disrespect of girls (or for black children, or disabled children), of emotional damage or cramped imagination in South African schools would raise questions about whether something we could call education was indeed taking place.

Selecting Capabilities for Gender Equality in Schooling

Taking all this into account, I now have a draft ideal-theoretical, multi-dimensional list taking up the specificity of gender equity in education, but with no attempt at a weighting or valuation of the various capabilities:

1. *Autonomy,* being able to have choices, having information on which to make choices, planning a life after school, independence, empowerment
2. *Knowledge,* of school subjects that are intrinsically interesting or instrumentally useful for post-school choices of study, paid work and a career; girls' access to all school subjects; access to powerful analytical knowledge, and including knowledge of girls' and women's lives; knowledge for critical thinking and for debating complex moral and social issues; knowledge from involvement in intrinsically interesting school societies, active inquiry; transformation of understanding; fair assessment/examination of knowledge gained
3. *Social relations,* the capability to be a friend, the capability to participate in a group for friendship and for learning, to be able to work with others to solve problems and tasks, being able to work

with others to form effective or good groups for learning and organizing life at school, being able to respond to human need, social belonging

4. *Respect and recognition,* self-confidence and self-esteem; respect for and from others; being treated with dignity; not being diminished or devalued because of one's gender, social class, religion, or race; valuing other languages, other religions, and spiritual practice and human diversity; showing imaginative empathy, compassion, fairness, and generosity; listening to and considering other persons' points of view in dialogue and debate in and out of class in school; being able to act inclusively

5. *Aspiration,* motivation to learn and succeed, to have a better life, to hope

6. *Voice,* for participation in learning, for speaking out, not being silenced through pedagogy or power relations or harassment, or excluded from curriculum, being active in the acquisition of knowledge

7. *Bodily integrity and bodily health,* not to be subjected to any form of harassment at school by peers or teachers, generally being safe at school, making own choices about sexual relationships, being able to be free from sexually transmitted diseases, being involved in sporting activities

8. *Emotional integrity and emotions,* not being subject to fear, which diminishes learning, either from physical punishment or verbal attacks; developing emotions and imagination for understanding, empathy, awareness, and discernment

Capability and Curriculum

But in education we also need to think about which additional theories we might draw on to further our understanding of learning, identity formation, and the construction of adapted preferences. (Also see Terzi, chapter 2, for a philosophically grounded list of basic education capability.) Theories about the school curriculum help us understand the capability of knowledge in more depth. Thomson (1999) points out that the school curriculum is a political arrangement, a selection from knowledge and a view on whose and what knowledge counts or is excluded or marginalized. "Students," as Thomson argues, "form their understanding of the world and their identities at least in part through the knowledges and narratives available to them in the curriculum" (1999, p. 11). Similarly, feminist educationalists such as Paechter (2003) have pointed to the relationship between gender, knowledge, and power in the school curriculum that works to girls' disadvantge. At issue is that the capability approach needs to keep in mind how knowledge and power work in schools to produce and reproduce inequalities, and to draw on educational theories that will provide ways of understanding such education effects. As Paechter (2000) points out, we need to devise strategies that genuinely allow girls and young women access to high-status knowledge and the power that accompanies it.

Having identified "knowledge" as a core education capability, we still need to ask how this is acquired, and for this we need additional theories about which pedagogic practices are most likely to result in girls developing valued capabilities. Freire (1972, p. 58) highlights the importance of how knowledge is mediated by teachers when he criticizes what he describes as "banking education" in which a teacher deposits knowledge into the blank and empty vessels of his or her passive students. He writes that knowledge instead should be a process of active inquiry: "Knowledge emerges only through invention and re-invention, through the restless, impatient, continuing, hopeful inquiry men [sic] pursue in the world, with the world and with each other." At issue is that in the pedagogical relationship produced between teacher and pupils, there is the possibility to enhance agency as also the possibility to deny agency. We need to have the means to know and understand the difference and the impact on capability. In order to do this we arguably need both capability theorizing and additional theorizing about learning and about identity formation.

In an innovative and creative approach in Australia, Reid (2005) has argued for a capability-based curriculum in Australian schools. Here he has in mind the foregrounding of a rich set of capabilities as the key outcomes of schooling. The acquisition of knowledge, which is usually the focus of curriculum outcomes, is then mediated through and by the fostering of capabilities. He outlines capabilities therefore that include what he describes as "knowledge work," and also go beyond curriculum-as knowledge work. Reid adds: innovation and design, productive social relationships, active participation, intercultural understandings, interdependence and sustainability, understanding self, ethics and values, and communication and multiliteracies to his list (see Reid 2005, p. 55 for a detailed description on each capability).

One could argue that it would be worthwhile to consider adding, or integrating, capabilities from his list not present in my own. Of course, what Reid's list, Terzi's list (chapter 2), and my own list all point to is the need for debate in particular contexts, debate that might begin with one or more lists, or that might begin somewhere else, say, with core educational values and principles. None of these lists make claims for universality; rather they make claims for the capability approach in education where the focus on capability outcomes is seen as contributing to social justice.

What this and other lists further underline is the multidimensionality of the capability approach—arguably all dimensions on this list are important for the quality of girls' education in South Africa. It serves as an ideal list that may need revising in nonideal practical contexts but the multidimensionality of which should not be sacrificed. I argue also that for my evolving education list, that to reduce the development of any one capability is to reduce the development also of others. All the capabilities on my provisional list are valuable for girls' education; if any one were not present in South African schools, then we would have cause to question the gender e/quality of education in those schools.

The lens of capability directs our attention to any sources of unfreedom that might constrain genuine choices and how diverse individuals are affected. For example, violence and harassment of female pupils by their male peers and by teachers continues to be endemic in large numbers of black schools so that many schools cannot be a place of substantive freedom, nor easily a place of "active, empowered capability" (Unterhalter 2003, p. 16). It then follows that while all children in South Africa have the right to be educated, they do not all have the capability to participate in what we understand to be education, that is, a process that enhances agency and well-being. Using the benchmark of capability equality, things are not going well for many South African students in school, and girls are particularly badly affected. We can argue this with greater confidence because we have begun to formulate a list of education capabilities, however revisable, that defines what it is we are assessing in relation to gender equality and social justice in education. In other words, we have a conception of what counts from the capability approach—agency, well-being, human dignity—and this counts as much or more for compulsory education as a public responsibility. We might go so far as to argue that all girls are not truly able to exercise the capability to participate in education or develop their capabilities for the wider benefits of learning if any one of the minimal constitutive capabilities on my list was absent or not being developed.

We might then introduce questions such as who has the power to develop these capabilities, and who has not? Put simply, in the context of girls' schooling, which girls and how? We might wish to check (measure) how successful girls are in bringing about what they are trying to achieve. Finally, if there is unevenness, patchiness, and inequality in girls' well-being freedom and agency freedom, we must then ask pedagogical, ethical, and policy questions about the society in which some girls can promote all their ends while others face barriers, whether of social class, race, gender, culture, or disability. At issue here is that interpersonal variations as well as individual capability must be considered.

Debating the List Publicly

To conclude, schools contribute, for most people quite substantially, to the formation of their capabilities to function. Ideally schools ought to equip girls with the capabilities to pursue opportunities they value. How valued and valuable opportunities and capabilities are distributed through formal education and to whom, and how this maps over structures of race, gender, class able-bodiedness, religion, and so on is then a matter of social justice in education and hence for education policy. The method by which a list of education capabilities comes to be produced is as significant as the contents of that list. It is for this reason that my final step emphasizes public debate and discussion, and why throughout the chapter I have been at pains to stress the ideal-theoretical nature of my own list. The method of selecting capabilities is imperfectly participatory, notwithstanding the girls' empirical voices. Yet, this should not detract from engaging the debate on the possibilities of a capability-based approach for gender equality policy making and evaluation in education.

This is a potentially radical approach. If we add to it the four key dimensions that Baker et al. (2004) identify for equality in education: resources, respect and recognition; power; and love, care, and solidarity, and their fifth dimension for equality, that of work and learning, we would have a powerful grid of dimensions and capabilities for producing and evaluating gender and social justice in education.

Acknowledgments

The research on which this paper is based is funded by the National Research Foundation (NRF) in South Africa under the project title "Pedagogy, Identity and Social Justice," Project No. NRF 4878. The financial support of the NRF is gratefully acknowledged. An earlier version of this chapter also appears in the *Journal of Education Policy* 21, no. 2, 2006.

Notes

1. Volunteer year 10 girls from four different schools in Cape Town were interviewed over a three-week period in August 2003. The schools were selected for their different histories and socio economic intake. Access was negotiated by writing to the Western Cape Education Department and thence to the head teacher of each school in the first instance, and then producing an explanatory leaflet for year 10 girls to enable them to decide whether or not they wished to participate. Before the interviews I held a meeting with the group of volunteers at each school to explain the project in person. Interviews were conducted with

pairs of girls, with each interview lasting around one and a half hours. Prior to the interview each pair of girls was given a disposable camera to take a set of photographs of their lives in and out of school. These formed the basis of the discussions with myself. All the girls' names mentioned are pseudonyms.

2. School A: 14 girls interviewed, 100% African pupils; majority working-class backgrounds; teachers 95% African; and school head African female. School B (all-girls school): 12 girls interviewed, 8 mixed race, 3 African, 1 white; pupils 80% mixed race, 10% African, 10% white; mostly middle class and lower middle class, small number of working-class backgrounds; teachers 90% white and school head white female. School C: 4 girls interviewed, all white; pupils 60% white, 30% mixed race, 10% African; middle-class backgrounds; teachers 90% white and school head white male. School D: 10 girls interviewed, all mixed race; pupils 80% mixed race, 10% African, 10% white; lower- and middle-class backgrounds; teachers 90% white and school head white male.

3. It was the requirement that African children study through the medium of the Afrikaans language, the "language of the oppressor", that ignited long simmering frustrations with bantu education on 16 June 1976 in Soweto, a turning point in the history of struggles against apartheid in South Africa.

References

Alkire, Sabina. 2002. *Valuing freedoms: Sen's capability approach and poverty reduction* Oxford: Oxford University Press.

Appadurai, Arjun. 2004. The capacity to aspire: Culture and the terms of recognition. In *Culture and Public Action*, edited by V. Rao and M. Walton. Stanford: Stanford University Press.

Baker, J. Kathleen Lynch, S. Cantillon, and J. Walsh. 2004. *Equality: From theory to action*. Houndmills: Palgrave Macmillan.

Department of Education (DOE). 2003. *Education statistics in South Africa: At a glance in 2001*. Pretoria: Department of Education.

Edwards, Richard, Stewart Ranson, and Michael Strain. 2002. Reflexivity: Towards a theory of lifelong learning. *International Journal of Lifelong Education* 21 (6): 525–536.

Freire, Paulo. 1972. *Pedagogy of the oppressed*. London: Sheed and Ward.

Gallacher, Jim, Beth Crossan, John Field, and B. Merrill. 2002. Learning careers and the social space; exploring the fragile identities of adult returners in the new further education. *International Journal of Lifelong Education* 21: 493–509.

Hage, G. 2001. The shrinking society: Ethics and hope in the era of global capitalism. *Australian Financial Review*, September 7.

Meerkotter, Dirk 1998. The state of schooling in South Africa and the introduction of Curriculum 2005. In *Vision and reality: Changing education and training in South Africa*, edited by W. Morrow and K. King. Cape Town: University of Cape Town Press.

Narayan, Deepa, and Patti Petesch. 2002. *Voices of the poor from many lands*. New York: World Bank and Oxford University Press.

Nussbaum, Martha C. 2000. *Women and human development: The capabilities approach*. Cambridge: Cambridge University Press.

———. 2003. Capabilities as fundamental entitlements: Sen and social justice. *Feminist Economics* 9 (2–3): 33–59.

Paechter, Carrie. 2000. *Changing school subjects: Power, gender and curriculum*. Buckingham: Open University Press.

———. 2003. Gender equality and curriculum change: What can we learn from historics in Western Europe, the USA and Australia. Paper read at Beyond Access seminar: Curriculum for gender equality and quality basic education in schools, September, University of London.

Ranson, Stewart. 1998. *Inside the learning society*. London: Cassell.

Reid, Alan. 2005. *Rethinking national curriculum collaboration: Towards an Australian curriculum*. DEST, Commonwealth of Australia, Canberra Available from http://www.dest.gov.au/research/publications/national_curriculum/default.htm (accessed February 1, 2006).

Robeyns, Ingrid. 2003. *Sen's capability approach and gender inequality: Selecting relevant capabilities* Available from http://www.st-edmunds.cam.ac.uk/vhi/nussbaum/papers/robeyns.pdf (accessed April 1, 2005).

Sen, Amartya. 1992. *Inequality re-examined*. Oxford: Oxford University Press.

———. 1999. *Development as Freedom*. Oxford: Oxford University Press.

———. 2004. What Indians taught China. *New York Review of Books* 19:61–66.

South Africa. 1996. *National Education Policy Act*. Pretoria: Government Printer.

Thomson, Patricia. 1999. How doing justice got boxed in: A cautionary curriculum tale for policy activist. In *Contesting the curriculum*, edited by B. Johnson and A. Reid. Sydney: Social Science Press.

Unterhalter, Elaine. 2003. The capabilities approach and gendered education: An examination of South African complexities. *Theory and Research in Education* 1 (1): 7–22.

Walker, Melanie. 1997. Women in the academy: Ambiguity and complexity in a South African university. *Gender and Education* 9 (3): 365–381.

———. 2005. *Higher education pedagogies: A capabilities approach*. Maidenhead: SRHE/Open University Press and McGraw-Hill.

———. 2006. *Higher education pedagogies: A capabilities approach*. Maidenhead: SRHE/Open University Press.

Wolpe, AnnMarie, O. Quinlan, and L. Martinez. 1997. *Gender equity in education: A report by the Gender Equity Task Team*. Pretoria: National Department of Education.

Wynn, J. 2004. Youth transitions to work and further education in Australia. Paper read at the annual meeting of the American Educational Research Association, April, San Diego.

10

Children's Valued Capabilities[1]

Mario Biggeri

Overview

This chapter considers the perceptions that children themselves hold regarding the relevance of education for their own well-being. The human development of children can be regarded as "an expansion of capabilities" or of "positive freedoms." Capabilities, choices, and conditions during childhood and adolescence crucially affect children's position and capabilities as adults. Deficiencies in important capabilities during childhood not only reduce the well-being of those suffering from the deficiencies, but may also have larger societal implications. Results from field studies carried out in Italy, India, and Uganda, which located children at the center of a bottom-up strategy for understanding the relevant dimensions of children's well-being, are reported. In relation to democratic dialogue about selecting capabilities, it is argued that children are capable of understanding and contributing thoughtful opinions. The overall concern is to demonstrate what children think they should be able to do and be, that is, their valued capabilities. It considers that an operationalization of the capability approach has to deal with the issue of defining a list of relevant capabilities, although this need not have a universal character.

An Operationalization of the Capability Approach

Practitioners of the capability approach regard education as a basic capability, irrespective of their stance on spelling out a complete definition of well-being in the space of capabilities. But what is the perception that children themselves have about the relevance of education for their own well-being? In taking up this question, I respond empirically to the observation that an operationalization of the capability approach has to deal with the issue of defining a list of relevant capabilities, although this need not have a universal character (Nussbaum 2003; Robeyns 2003, 2005). As suggested by Sen (2005), the open and public validation of such a list is a necessary condition to confer legitimacy on it, and such validation entails a thorough exercise of public scrutiny and debate, that is, an effort of deliberative democracy.

In relation to such a democratic process, there is evidence of many issues that even very small children are capable of understanding and to which they can contribute thoughtful opinions (Lansdown 2001). Article 12 of the Convention on the Rights of the Child (CRC) (UN 1989) is a milestone for the advocacy of children's participation, arguing as it does for the rights of the child to freedom of association and to peaceful assembly. It argues that "there is no lower age limit imposed on the exercise of the right to participate. It extends therefore to any child who has a view on a matter of concern to them" (Lansdown 2001, p. 2). Additionally, the United Nations report, *A World Fit for Children* (UNICEF 2002), argues that children are the subjects of rights and participants in actions affecting them. Consequently, the innovative feature of the research that will be presented in this chapter is to regard children as active participants in the debate around their own well-being.

The human development of children can be regarded as "an expansion of capabilities" or of "positive freedoms." Capabilities, choices, and conditions during childhood and adolescence crucially affect children's position and capabilities as adults (Sen 1999). Deficiencies in important capabilities during childhood not only reduce the well-being of those suffering from them, but may also have larger societal implications (Klasen 2001, p. 422). The aim of this paper is to explore how children value their capabilities and, in particular, the relevance they attribute to education. To this end, the results of three field studies carried out in Italy, India, and Uganda are reported and described in detail. These studies were meant to establish children at the center of a bottom-up strategy for understanding the relevant dimensions of children's well-being. As already mentioned, we were concerned with what children think their valued capabilities are.

Such a bottom-up approach to the definition of the relevant set of capabilities does not impose an external value judgment on the analysis of well-being; at the same time, it provides a standard framework for the operationalization of Sen's approach (Biggeri et al. 2006a). In these three studies we let children establish which capabilities are relevant for them, keeping to a minimum any external interference. On the other hand, however, this effort has its own problems and complexities: expressed perceptions may not reflect children's real desires, preferences, and underlying values, as these could have been thwarted by adaptation to their living conditions. We attempted to deal with this issue by developing a process meant to be conducive to reflective reasoning around individual preferences, ideally detaching children from the constraints arising from personal experience.

Five Issues Related to Children's Capabilities

There are at least five issues related to children's capabilities that are worth recalling (Biggeri 2004; Biggeri et al. 2006a). The first observation concerns the fact that the child's capabilities are at least partially affected by the capability set and achieved functionings (as also by their means, i.e., assets, disposable income) of their parents, as an outcome of a cumulative path-dependent process that can involve different generations of human beings (Biggeri 2004). The second observation is that the possibility of converting capabilities into functionings depends also on parents', guardians', and teachers' decisions, implying that the child's conversion factors are subject to further constraints. On the one hand, parents need to respect children's desires and freedoms, but on the other they have to assist children to expand or acquire further capabilities, even though this may need to be done against children's willingness (Biggeri et al. 2006a). This can become relevant to an education capability where parents and tutors can be inspired by different motivations and they can be either autonomy supportive (e.g., giving an internal frame of reference, providing meaningful rationale, allowing choices, encouraging self-perspective) or just controlling (e.g., pressure to behave in specific ways). These two aspects can be in conflict since the child is not a passive actor, especially as age increases. Therefore, the degree of autonomy is relevant in the process of choice.

A third relevant aspect is connected to the relationship between different capabilities and functionings. The fact that education is a basic capability with an intrinsic value means that it can be instrumental for other capabilities (see Terzi chapter 2 and Vaughan chapter 6). Indeed, it may affect the current and perspective capabilities of the child. The fourth aspect concerns the life cycle and the importance of age in defining the relevance of a capability. This means that a careful timing of interventions is required for a child's well-being, including different types of education objectives according to the age and the maturity of the child.

The last issue concerns the role of children in building-up the future society and its constraints. Children, from this point of view, can be considered as a vehicle of change; once they reach adulthood they can contribute to shaping future conversion factors.

As we have argued elsewhere (Biggeri et al. 2006a), there are numerous reasons why policymakers should place higher priority on children's capabilities. A key component of such a child-oriented approach is a policy that ensures universal access to education (UNESCO 2001), with attention to the quality of education.

Background to the Three Research Studies

In the three studies—carried out in Italy, India, and Uganda—children were invited to interact and express their opinions on the most relevant issues related to their childhood and adolescence. In order to let the children establish their priorities and better understand their capabilities we planned ad hoc surveys accompanied by focus group discussions (FGDs) and case studies. The surveys were based on a common survey-based method that employed a questionnaire as a means of stimulating the process of thinking and participation.[2] Each child was asked to think about their capabilities and how relevant these were for them as an individual and as a group of human beings. The questionnaire was conceived to implement a bottom-up process whereby the children were encouraged to conceptualize and attribute value to specific children's capabilities.

Two opportunities for data collection were provided by international conferences in which children as participants and delegates were asked to analyze relevant issues regarding a child's life.[3] These conferences—organized by the Global March against Child Labour (GMACL) and other grassroots associations—were the first Children's World Congress on Child Labour (CWCCL) (Florence, Italy, in May 2004) and the second Children's World Congress on Child Labour and Education (CWCCLE) (Delhi, India, in September 2005). During the conferences a full census of the child delegates was conducted.[4]

The child delegates, although they had been elected by other children through a democratic consultation process,[5] were not assumed to be representative of all the world's children. Nonetheless, we argue that this does not invalidate their contribution (Lansdown 2001). These samples were both selective and of high quality, not only by virtue of having the important characteristic of being delegates of other children, but also because these children had acquired a high level of consciousness through their life experiences, especially through their participation in NGO activities. Indeed, both the research group and the conference organizers believed that the child delegates who took part in the congress—considering their life experiences as former child laborers and activists—could understand better than adults a child's wishes concerning how their life should progress and what would constitute a more equal and humane world for them (Biggeri et al. 2006a; GMAACL 2004; Cutillo 2004).

The third study was purposely not connected to a conference and aimed to interview children during their daily life. The research was carried out in Kampala, Uganda, in March/April 2005. Although the research was centered on street children[6] we decided to survey three categories of children:

1. Children of the street. These are the so-called full-time street children: the street is their home
2. Rehabilitated children who live in NGOs or other associations for children of the street. These are children who have been rehabilitated, in centers or NGOs or similar institutions, where they are accommodated and where they receive food, medical care, and vocational training
3. A control group made up of children of the same age and from the same localities who had never experienced "the street"

The distinction between these three categories allowed us to understand better how children with different experiences of life conceptualize relevant capabilities and value them.

Methodology

In each of the three studies we implemented an ad hoc survey based on a core questionnaire designed by the research group. The questionnaire was divided into different sections including a core capabilities section.[7] The questionnaire was conceived bearing in mind the importance of maintaining the children's full attention and participation.[8] In the capability section, each child undertook a process of reflection to help them to separate sufficiently from their own specific life experience. We decided to validate the relevance of a capability if two conditions are satisfied (i) if at least one child delegate identified it spontaneously, and (ii) if it was considered by the majority of child delegates as either important or very important. In the case of the CWCCL and the CWCCLE a full census of the delegates was conducted, while in the case of the research on street children in Kampala, three sample designs were planned[9] according to the three categories of children.

Although street children are a relevant and visible phenomenon in Kampala, they are nevertheless still invisible in official statistics. Considering the scarce quantitative information available (not enough to use random sampling directly), we decided to conduct an ad hoc survey with a multistage sample design (for details see Mariani et al. 2005). For these reasons the sampling design had to follow a specific method, taking into account qualitative information available about children of the street in Kampala and about which institutions were working fully to rehabilitate them. Data collection was thus based on a two-stage sample design. The first stage was the identification of the areas to conduct the interviews with the children of the street and of the more reliable NGOs for rehabilitated children. We were able, after a lengthy process, to select and identify a list of 12 NGOs. The second stage for both sample surveys was based on a random selection.[10] The *dimension of the samples* was chosen a priori to include at least 50 children for each category. The sample was sufficiently large to generate reliable data for statistical testing. The outcome of this sampling method for children of the street and for rehabilitated children can be considered a good representation of the respective statistical population in Kampala. In order to have a category for comparison, a control group of nonstreet children in the same geographical area was included in the sampling design. The control group consisted of children in the same neighborhood who had never been street children. Five state schools in the neighborhood were randomly selected. Within each of the schools children for a control group were randomly selected.

Children's Valued Capabilities

Educational Achievements

During the CWCCL (Italy), 104 out of the total 105 child delegates were interviewed. The delegates were from 45 countries, with a slightly higher percentage from developing countries. More than half of the child delegates were ex-child workers or laborers. Most of them (65%) were aged between 15 and 17 years, while the rest were children between 11 and 14 years.[11] Female delegates were 59% of the total. Although at the time of the interviews 94.2% of the children were attending school, some of them had had to drop out at some stage because of their household's economic hardship and start working. Considering all children delegates, 39.4% of them received nonformal education and of these 28.6% received it from people in the family (parents, sisters, brothers, uncles, etc.), 40.6% from NGOs, and 9.5% from missionaries and religious institutions.

During the CWCCLE (India), we interviewed 59 delegates. The delegates were from 21 countries, with most of them coming from developing countries, and more than half of them were ex-workers. Of the delegates interviewed 47.5% were aged between 9 and 14 years, while 52.5% were between 15 and 17 years old. Female delegates interviewed were 54.2% of the total. Although at the time of the interviews 83.1% of the children were attending school, 24% had had to drop out because of their household's economic hardship (57%) or because they had had to start to work (28.6%). Then, 50.8% of all children received nonformal education and of these 66.6% received it from NGOs.

For the research in Kampala, 158 children were interviewed according to the three samples: 53 children of the street, 55 rehabilitated children, and 50 as a control group. Of the 158 children interviewed, 69% were male and 31% female. According to the age category 16.6% were between 8 and 11 years old, 44.5% were between 12 and 14 years old, and 38.9% were between 15 and 17 years old. Concerning achieved functioning on education within the three categories, 86% of children of the street considered their education as insufficient; this share is reduced to 32% with rehabilitated children, and to 12.5% for the control group. This is confirmed by the fact that 13.1% of children of the street had received no education at all, 52.2% had reached only level P4 (four years) in the primary school, and only 11.3% had completed primary level. At the time of the interview only 6.4% of these children were attending school (formal or informal) and 93.6% had stopped, or even had not started schooling mainly for the following reasons: economic hardship 64%, the civil conflict in Uganda 9.1%, they were mistreated at home 10%, or they had left home 4%. Only 11.3% of the children of the street received informal education (16.7% from parents, 33.3% from sisters or brothers, 50% from NGOs, and 33.3% from missionaries or religious institutions). The rehabilitated children were in a better position since all of them had some primary education, and 56.4% were attending school (formal or informal). In the past 90.9% of them had had to drop out, most of them for economic hardship (78%), because they were mistreated at home (6%), abandoned (4%), or because they had left home (2%). In the control group, the share of children with at least some education reached 100%. Although the share of those attending school is 100%, as many as 22% of them had had to drop out for a period. As much as 81% of the drop out was attributed to household economic hardship, showing that the dropping out is a strategy of the household in difficult periods. It is interesting to note that 62% of the control group children received informal education (65.8% from parents, sisters or brothers, or uncles, and only 16.1% from NGOs and from missionaries or religious institutions).

Children's Valued Capabilities

After this analysis of educational achievement, which highlights the difference among children in education as an achieved functioning, we move to an analysis of the core section on capabilities. One of the other main results of the research is also connected to the methodology applied, which uses a survey-based technique to stimulate the child, through a participatory process of reflection, to conceptualize the child's relevant capabilities.

The first question to the children was an open one and is fundamental since they were asked to indicate: "What are the most important opportunities a child should have during her/his life?". The child, at this point of the interview, was not aware of capability as a concept, nor of the list of child capabilities that would be used to codify the answers. This question, as already noted, thus allowed the researcher to identify which capabilities were considered relevant by the children without any interference or suggestions. If the child mentioned a new capability (that is not included in the codified set, see later) it was then recorded; if the child mentioned one of those in the list it was marked (in cases where interpretation was difficult, the interviewer added it as a "new" capability, which was inserted and integrated in the following questions regarding capabilities). In this way each child could interact and participate directly in the formulation of the questionnaire (Biggeri et al. 2006a).

The interviewers worked with a codified list of 14 capabilities (Biggeri, 2004):

1. *Life and physical health:* being able to be physically healthy and enjoy a life of normal length
2. *Love and care:* being able to love and be loved by those who care for us and being able to be protected*
3. *Mental well-being:* being able to be mentally healthy
4. *Bodily integrity and safety:* being able to be protected from violence of any sort*
5. *Social relations:* being able to enjoy social networks and to give and receive social support*
6. *Participation:* being able to participate in public and social life and to have a fair share of influence and being able to receive objective information*
7. *Education:* being able to be educated
8. *Freedom from economic and noneconomic exploitation:* being able to be protected from economic and noneconomic exploitation
9. *Shelter and environment:* being able to be sheltered and to live in a safe and pleasant environment
10. *Leisure activities:* being able to engage in leisure activities
11. *Respect:* being able to be respected and treated with dignity
12. *Religion and identity:* being able to choose to live or not according to a religion (including peace with God, or the gods) and identity*
13. *Time-autonomy:* being able to exercise autonomy in allocating one's time and undertake projects*
14. *Mobility:* being able to be mobile*

It is significant to note that some capabilities can be more relevant as age increases. Indeed, the presence of an asterisk (*) in the list above indicates that, on the basis of previous studies (Lansdown 2001) and field experience, up to a level that is appropriate, given the age and maturity of the child, the relevance of the capability may vary [whether the child is in "early" childhood (from 0–5 years old), childhood (6–10), "early" adolescence (11–14), or adolescence (15–17)]. For instance, it is possible that different ages may attach different importance to each of the above-mentioned capabilities, while the complete list of capabilities may be fully enjoyed from the process aspect of freedom only by the older categories of children (Biggeri et al. 2006a).

During the survey, some capabilities were added by the interviewers according to the replies of the child delegates, but all of them were in any case reflected in those already codified. During the interview process, around 20 children conceptualized at least four possible categories of capabilities that were not directly codified by the interviewers and were therefore added at the end of the list. However, during the analysis of the questionnaires, the four categories were absorbed into the original codified list. For example, "national identity" was inserted under the "religion and cultural identity" category. The same occurred in India and in Uganda (see Biggeri et al. 2006b).

The results of the list of capabilities individually mentioned by at least one child are reported in table 10.1 for the two conferences and in table 10.2 for Uganda. These capabilities, as we explain later, were further legitimated by the fact that all of them were considered as important or very important by the majority of the children.

It is important to highlight that some capabilities—education, love and care, leisure activities, and life and physical health—were more frequently identified than others . We would like to point out the relevance of education to over 88% of the children at the two conferences. In the case of the CWCCL, having a larger number interviews, children can be grouped in different categories according to gender, their place of origin (developing and developed countries), and their past experience as workers or not. The results (table 10.1) show that all groups conceptualized the same categories of capabilities, suggesting that in our sample, children's points of view across cultural and economic divides are broadly similar.

Furthermore, an FGD carried out with eight South Asian child delegates (Nepal 3, Pakistan 2, and India 3) aged between 13 and 17 during the CWCCL emphasized the importance of education to these children (Biggeri et al. 2006a). In other words, children in the case studies (Menchini 2006) and the FGD are aware of the importance of education as an instrumentally valuable capability for their present and future well-being.

In Uganda (158 children) the results are very similar: 89% of the children mentioned education as a relevant capability, followed by life and physical health, and love and care. Analyzing the responses by the three categories

Table 10.1 Percent of child delegates who identified the capabilities at the CWCCL and at the CWCCLE*

Relevant capabilities	Total	Age group		Sex		Country of origin		Ever worked		CWCCLE*
		11–14	15–17	Female	Male	Developed	Developing	No	Yes	
1) Life and physical health	29.8	22.2	33.8	31.1	27.9	36.7	23.6	37.8	23.7	35.6
2) Love and care	48.1	58.3	42.6	52.5	41.9	53.1	43.6	53.3	44.1	40.7
3) Mental well-being	5.8	11.1	2.9	4.9	7.0	6.1	5.5	6.7	5.1	8.5
4) Bodily integrity and safety	17.3	13.9	19.1	16.4	18.6	28.6	7.3	24.4	11.9	3.4
5) Social relations	8.7	8.3	8.8	9.8	7.0	8.2	9.1	13.3	5.1	8.5
6) Participation/information	13.5	5.6	17.6	18.0	7.0	14.3	12.7	15.6	11.9	25.4
7) Education	89.4	94.4	86.8	88.5	90.7	89.8	89.1	88.9	89.8	88.1
8) Freedom from economic and noneconomic exploitation	11.5	8.3	13.2	8.2	16.3	16.3	7.3	8.9	13.6	11.9
9) Shelter and environment	13.5	11.1	14.7	14.8	11.6	16.3	10.9	15.6	11.9	28.8
10) Leisure activities	34.6	47.2	27.9	37.7	30.2	30.6	38.2	37.8	32.2	18.6
11) Respect	12.5	13.9	11.8	11.5	14.0	10.2	14.5	6.7	16.9	6.8
12) Religion and identity	3.8	2.8	4.4	3.3	4.7	4.1	3.6	2.2	5.1	6.8
13) Time autonomy and undertaking of projects	11.5	8.3	13.2	16.4	4.7	14.3	9.1	13.3	10.2	1.7
14) Mobility	3.8	0.0	5.9	3.3	4.7	4.1	3.6	2.2	5.1	3.4

Question: "What are the most important opportunities a child should have during his/her life?"
Note for the interviewer: Do not read out; multiple answers allowed; add capabilities not present in the list of codified at the end
* Preliminary results
Source: Biggeri et al. (2006a)

Table 10.2   Percent of children who identified the capabilities at Kampala, Uganda

Relevant capabilities	Street children	Rehabilitated children	Control group
1) Life and physical health	83.0	87.3	78.0
2) Love and care	81.1	81.8	82.0
3) Mental well-being	30.2	14.5	10.0
4) Bodily integrity and safety	15.1	12.7	22.0
5) Social relations	26.4	14.5	28.0
6) Participation/information	15.1	10.9	14.0
7) Education	79.2	94.5	94.0
8) Freedom from economic and noneconomic exploitation	20.8	10.9	18.0
9) Shelter and environment	34.0	30.9	30.0
10) Leisure activities	28.3	29.1	48.0
11) Respect	11.3	3.6	6.0
12) Religion and identity	3.8	5.5	24.0
13) Time autonomy and undertaking of projects	9.4	3.6	12.0
14) Mobility	11.3	10.9	8.0

Question: "What are the most important opportunities a child should have during his/her life?"
Note for the interviewer: Do not read out; multiple answers allowed; add capabilities not present in the list of codified at the end
Source: Mariani et al. 2005

of children, we found that for the rehabilitated children and for the control group children, education is rated very highly: 94.5% and 94%. In the case of children of the street education is rated slightly lower than the other two capabilities mentioned above, but it still has a very high share of 79%.

We included a question where each child had to concentrate on individual experience (achieved functionings) for each dimension identified, and then another question to conceptualize the relevance of the same capability for children as a group. This sequencing in the process of thinking should enable children to be at least partially detached from their own life experience.

The results for these two questions are very interesting: (i) the level of the achieved functionings varies according to a child's life experience and according to the categories of children, especially in Uganda; (ii) children value each single capability identified in tables 10.1 and 10.2 as important or very important, with percentages well over the majority.

The last question of the section was intended to select the most relevant capabilities and to "value" them. Without providing a complete ordering of all capabilities, we asked the children to select the three most relevant.[12] Among them, the first three mentioned more often among the most valued are education, love and care, life and physical health (table 10.3).

Table 10.3 Results of core question on capabilities at the CWCCL and at the CWCCLE* (percent): The three most relevant capabilities

Relevant capabilities		CWCCL								CWCCLE*
	Total	Age group		Sex		Country of origin		Ever worked		
		11–14	15–17	Female	Male	Developed	Developing	No	Yes	
1) Life and physical health	34.6	33.3	35.3	32.8	37.2	38.8	30.9	37.8	32.2	38.0
2) Love and care	51.9	52.8	51.5	50.8	53.5	59.2	45.5	51.1	52.5	49.4
3) Mental well-being	9.6	11.1	8.8	14.8	2.3	18.4	1.8	13.3	6.8	13.3
4) Bodily integrity and safety	5.8	5.6	5.9	3.3	9.3	8.2	3.6	8.9	3.4	3.8
5) Social relations	3.8	5.6	2.9	3.3	4.7	4.1	3.6	2.2	5.1	1.9
6) Participation/information	18.3	8.3	23.5	23.0	11.6	22.4	14.5	15.6	20.3	22.8
7) Education	73.1	69.4	75.0	77.0	67.4	65.3	80.0	66.7	78.0	87.3
8) Freedom from economic and noneconomic exploitation	25.0	36.1	19.1	24.6	25.6	26.5	23.6	35.6	16.9	15.2
9) Shelter and environment	13.5	11.1	14.7	6.6	23.3	18.4	9.1	15.6	11.9	22.8
10) Leisure activities	24.0	33.3	19.1	21.3	27.9	8.2	38.2	17.8	28.8	11.4
11) Respect	11.5	13.9	10.3	14.8	7.0	10.2	12.7	13.3	10.2	13.3
12) Religion and identity	2.9	0.0	4.4	3.3	2.3	2.0	3.6	2.2	3.4	7.6
13) Time autonomy and undertaking of projects	9.6	11.1	8.8	6.6	14.0	10.2	9.1	11.1	8.5	7.6
14) Mobility	0.0	0.0	0.0	0.0	0.0	0.0	0.0	0.0	0.0	5.7

Question: "Among the aspects we discussed could you tell me which are the three most important opportunities a child should have during his/her life?"
* Preliminary results
Source: Biggeri et al. (2006a)

Table 10.4   Results of core question on capabilities at Kampala, Uganda (percent):
The three most relevant capabilities

Relevant capabilities	Street children	Rehabilitated children	Control group
1) Life and physical health	45.5	60.8	69.9
2) Love and care	51.7	34.2	57.5
3) Mental well-being	8.3	0.0	4.1
4) Bodily integrity and safety	4.1	1.9	4.1
5) Social relations	4.1	0.0	0.0
6) Participation/information	12.4	9.5	2.1
7) Education	68.3	89.2	84.2
8) Freedom from economic and noneconomic exploitation	10.3	0.0	2.1
9) Shelter and environment	37.2	53.2	39.0
10) Leisure activities	12.4	11.4	10.3
11) Respect	10.3	11.4	2.1
12) Religion and identity	8.3	11.4	4.1
13) Time autonomy and undertaking of projects	6.2	5.7	8.2
14) Mobility	12.4	7.6	6.2

Question: "Among the aspects we discussed could you tell me which are the three most important opportunities a street child should have during his/her life?"
Source: Mariani et al. 2005

Education stands out among the capabilities identified, since 73% of the children recalled it as one of the three most important capabilities in the CWCCL and 87% in the CWCCLE. In Kampala, the results are remarkable with education always recalled among the three most relevant capabilities (table 10.4). When the three categories of children are analyzed, the results underline that the children of the street mentioned among the three most important capabilities education (68.3%), love and care (51.7%), life and physical health (45.5%).

Conclusions

Deficiencies in important capabilities such as education during childhood reduce well-being even in the future (poverty as deprivation of capabilities) and have larger societal implications. Children in the three research studies were asked to express themselves on the most important issues related to their childhood and to identify a list of capabilities that were important to them. The results confirm that children, independently, from different countries and experiencing different circumstances are able to conceptualize relevant capabilities and that, in particular, they value education as one of the most important capabilities. Of course, education may also have a negative impact on other dimensions of well-being by affecting other capabilities. For instance, a child's life centered fully and excessively on education may reduce other important capabilities such as leisure activities, social relations, and participation, which are, according to children, important capabilities as well.

Nonetheless, I argue finally that policymakers should put all efforts and a higher priority on children's capabilities and on education. The goal of "quality" "education for all" children is an ethical, social, political, and moral imperative.

Acknowledgments

Our warm thanks to the GMACL and especially the chairperson, Kailash Satyarthi, and the director, Alok Vajpeyi. The author is grateful to the NGO Mani Tese and particularly to its president, Filippo Mannucci, and Mariarosa Cutillo for their full support. Special thanks to the research group of the University of Florence composed of Renato Libanora, Mariani Stefano, Leonardo Menchini, Simone Bertoli, and Rudolf Anich as well as the thematic group on children's capabilities of the Human Development and Capability Association (HDCA). Sincere thanks to Sabina Alkire, Nicolò Bellanca, Enrica Chiappero Martinetti, Flavio Comim, Giovanni Andrea Cornia, Santosh Mehrotra, David Parker, Aesa Pighini, Mozaffar Qizilbash, Ingrid Robeyns, Fabio Sani, Marta Santos Pais, and Franco Volpi for their useful suggestions in the three researches. The paper has benefited from the comments of Melanie Walker and Elaine Unterhalter and the participants at the Fifth International Conference on the Capability Approach, September 2005, UNESCO, Paris. The author is extremely grateful to the CWCCL and the CWCCLE child delegates (and their accompanying persons) since, fully understanding the importance of the research project, both collaborated extensively and contributed to its success. For the research in Italy, I am indebted to the PhD students in Politics and Economics of Developing Countries (University of Florence) for helping us in interviewing all the child delegates, and in particular to Simone Bertoli, Giovanni Canitano, Marika Macchi, Maria C. Paciello, Giorgio Ricchiuti, Elisa Ticci, Monica Bianchi, Chiara Mariotti, and Valeria Pecchioni. For the pilot survey in Nepal, I wish to thank Luca Biggeri and Sabindra Khadgi. I am extremely grateful to Rudolf Anich for his excellent support as coordinator of the field researches both in India and in Uganda. The precious help of Lameck Muwanga, Semakula Musoke Henry, Opolot Roberts Imongot, Alutia Moses Zorran Brudd, and the "baabas" in Uganda and of Avinash Singh and Barbara Russo in India is fully acknowledged. The author is very thankful to the children interviewed in Uganda.

Stefano Mariani carried out the database management and elaboration. The methodology section is written and attributed also to Mario Biggeri, Stefano Mariani, Renato Libanora, Leonardo Menchini, and Rudolf Anich. The author retains responsibility for the opinions expressed in the paper.

Notes

1. This chapter summarizes the results of two research projects. The first is "Children Establishing Their Priorities: Developing Bottom Up Strategies for Understanding Children's Well Being and Childhood, and their Impact on Research on Child Labour," a collaboration between the organizers of the First and Second Children's World Congress on Child Labour (held in Italy, 2004, and India, 2005), especially the Global March against Child Labour (GMACL), and the NGO Mani Tese and the PhD Course in Politics and Economics of Developing Countries at the University of Florence, Italy. The second is a study on the "Capability Deprivation of Street Children in Kampala (Uganda)" in March–April 2005 by the thematic group on "Children's Capabilities" of the Human Development and Capability Association (HDCA). Both the studies were carried out by a research group of the University of Florence composed of economists, development economists, statisticians, demographers, anthropologists, and psychologists. I acknowledge the support of the University of Florence and of the Fondazione Culturale Responsabilità Etica (the funds, which enabled us to cover some expenses, were donated by the family of Pia Paradossi, who was a much-loved member of Mani Tese Firenze).
2. According to Alkire (2002, p. 129), participation falls along an axis "between 'light' interaction in which participants supply information, to 'deep' participation in which the control of the decision-making and resources are in the hands of the participants." Our research method is closer to "light" participation (see Bertoli 2006).
3. It is important to establish clear principles and ground rules that guide events involving children's participation. Indeed, there is a danger that adults might use children to promote their goals. We compared this conference organization process with the criteria reported by Lansdown (2001, pp. 30–39) and, in terms of these criteria, we can confirm that the CWCCL was "owned" by the children and young people throughout the entire process and considered as a conference run by and for children, with adult support.
4. Please see note 3.
5. As reported by GMACL (2004), the participants were chosen, with a balance of girls and boys aged between 10 and 17 years, by the children themselves through a fair and democratic selection process during national and regional consultations intended to avoid any kind of discrimination. A common feature of the participants was the involvement in the cause of ending child labor, the promotion of universal and quality education, and the awareness of the rights of the child.

6. As far as we know, there is nothing in the literature that specifically addresses the issue of street children capabilities, with the exception of our study (Mariani et al. 2005), here reported, and the study in Mauritania of Ballet, Bhukuth, and Radja (2004).

7. "The questionnaire was divided into five different sections: i.) Personal Characteristics or Introductory Section; ii.) Education Section (formal/informal); iii.) Work Section; iv.) Capabilities Section (the core of the questionnaire); v.) General and Policy Section. The questionnaire consisted of a total of 20 items (or main questions) some of which are further divided into sub-questions (60 questions in total). As stated at the beginning of the questionnaire, all the information collected in the survey is strictly confidential and was used for statistical purposes only. A brief manual on the purpose of the research and on how to conduct the interviews was also prepared" (Biggeri et al. 2006a, p. 67). A section "i.)" was added to get information on street children so that the number of items become 25 and the number of questions 70.

8. Maintaining a code of ethics during the research process is essential as children have less power than adults. The researcher is responsible for making sure that the research will not be harmful to the children and that participation in research is voluntary (Boyden 1997; ILO-UNICEF 2000; Lewis and Lindsay 2000; Lansdown 2001; RWG-CL 2003; Laws and Mann 2004). The questionnaire, as well as the validity of a briefing manual, was tested in each survey. We then prepared one or two days' workshop for training the interviewers. The interviews were conducted directly with the children by the interviewers; only in a few cases was the help of interpreters needed.

9. Both quantitative (the ad hoc questionnaire) and qualitative methods were adopted for the research of street children in Kampala. In parallel with the ad hoc survey, other activities were carried out during the field research relating to and including children's life histories, children's mapping, children's drawings, and children's photos. For details on the qualitative methods and analysis see Anich (2006).

10. Another very important characteristic is that children of the street cannot be approached easily for interviewing. Therefore it was very important to find somebody who can act as a link between the research group and the street children. We therefore worked with the "baabas," former street children, now adults and fully reintegrated in society, who proved to be particularly responsible and were thus employed by different NGOs in order to identify, to sensitize, and to support the current street kids (Anich 2006). A small fraction of the children were selected through the snowballing technique since some children of the street brought us the first friend they met after the interview.

11. This imbalance was partially due to problems connected to immigration visas and permits, because the Italian government thought some of the children a security risk. (On this issue please see the children's final declaration, May 2004, sic., www.globalmarch.org).

12. "Among the aspects we discussed could you tell me which are the three most important opportunities a child should have during her/his life?". In Uganda it

refs only to street childern. Please, see figure 10.4. This is why only the first column in commented.

References

Alkire, Sabina. 2002. *Valuing freedoms: Sen's capability approach and poverty reduction.* Oxford: Oxford University Press.

Anich, Rudolf. 2006. *Bambini di strada. Indagine sociologica di un recente fenomeno urbano attraverso le parole e le immagini dei bambini di Kampala.* Tesi di laurea in Scienze Politiche, Università degli Studi di Firenze.

Ballet, Jérôme, Augendra Bhukuth, and Katia Radja. 2004. Capabilities, affective capital and development application to street child in Mauritania. Paper read at the Fourth International Conference on the Capability Approach: Enhancing Human Security, September 5–7, at Pavia, Italy.

Bertoli, Simone. 2006. The circumstantial contingency of desires and participatory definitions of well-being in the capability space (mimeo). In *Dottorato in politica ed economia dei paesi in via di sviluppo, working paper 3/3.* Università di Firenze.

Biggeri, Mario. 2004. *Capability approach and child well-being, Studi e discussioni, n 141, Dipartimento di Scienze Economiche.* Florence: Università degli Studi di Firenze.

Biggeri, Mario, Renato Libanora, Stefano Mariani, and Leonardo Menchini. 2006a Children conceptualizing their capabilities: Results of the survey during the first Children's World Congress on Child Labour. *Journal of Human Development,* 7 (1): 59–83.

Biggeri, Mario, Rudolf Anich, Stefano Mariani, and Renato Libanora. 2006b. Street children in Kampala: Understanding capabilities deprivation through a participatory approach. Paper read at the Sixth International Conference of the Human Development and Capability Association, August 29–September 1, Groningen, the Netherlands.

Boyden, Jo. 1997. A comparative perspective on the globalization of childhood. In *Constructing and reconstruction childhood, contemporary issues in the sociological study of childhood,* edited by A. James and A. Prout. London: Routledge, Falmer.

Cutillo, Mariarosa. 2004. Un impegno di giustizia: la lotta allo sfruttamento dell'-infanzia e per l'accesso all'istruzione universale. *Sistema Previdenza, Bimestrale di Informazione dell'Istituto Nazionale della Previdenza Sociale* 21 4/5.

GMCAL. 2004. *Children's world congress on child labour: Narrative report.* New Delhi: Global March against Child Labour, International Secretariat.

ILO-UNICEF. 2000. Investigating child labour, guidelines for rapid assessment: A field manual (mimeo). Geneva: SIMPOC.

Klasen, Stephan. 2001. Social exclusion, children and education: Implications of a rights-based approach. *European Societies* 3 (4): 413–445.

Lansdown, Gerison. 2001. *Promoting children's participation in democratic decision-making.* Florence: Innocenti Insight, UNICEF Innocenti Research Centre.

Laws, Sophie, and Gillain Mann. 2004. *So you want to involve children in research?* London: Save the Children.

Lewis, Ann, and Geoff Lindsay, eds. 2000. *Researching children's perspectives.* Buckingham: Open University Press.

Mariani, Stefano, Renato Libanora, Rudolf Anich, and Mario Biggeri. 2005. A methodology for understanding capabilities and for individuating social policies: An application to street children in Kampala. Paper read at the Fifth International conference of the Human Development and Capability Association, September 11–14, at UNESCO, Paris, France.

Menchini, Leonardo. 2006. Case studies (mimeo). In *Children establishing their priorities,* edited by M. Biggeri. Milan: Report for Mani Tese Italia (mimeo).

Nussbaum, Martha C. 2003. Capabilities as fundamental entitlements: Sen and social justice. *Feminist Economics* 9 (2–3): 33–59.

Robeyns, Ingrid 2003. Sen's capability approach and gender inequality: Selecting relevant capabilities. *Feminist Economics* 9 (2–3): 61–91.

———. 2005. The capability approach: A theoretical survey. *Journal of Human Development* 6 (1): 93–114.

RWG-CL. 2003. *Handbook for action-oriented research on the worst forms of child labour including trafficking in children.* Bangkok: Regional Working Group on Child Labour.

Sen, Amartya. 1999. Investing in early childhood: Its role in development. Paper read at the Conference on Breaking the Poverty Cycle. Investing in Early Childhood, at Washington, D.C.

———. 2005. Human rights and capabilities. *Journal of Human Development* 6 (2): 151–166.

UN. 1989. *Convention on the rights of the child.* New York: United Nations.

UNESCO. 2001. *Education for All and children who are excluded, World Education Forum, Dakar, Senegal, April 26–28, 2000.* Paris, France: UNESCO.

UNICEF. *A world fit for children (A/S-27/19/Rev.1)* UNICEF 2002. Available from http://www.unicef.org/specialsession/documentation/documents/A-S27-19-Rev1E-annex.pdf.

11

The Capability Approach and Women's Economic Security: Access to Higher Education under Welfare Reform

Luisa S. Deprez and Sandra S. Butler

Education is the single most consistent and powerful instrument for the advancement of an individual and a people.

(*Johnnetta B. Cole*)

In modern industrialized nations, the relationship between the two fundamental functions of higher education is both profound and inseparable: for the society, it maintains and transmits culture, values and norms from one generation to another; for the individual, it helps one overcome disadvantage and gain greater control of one's life. All pathways point to its promise, prospect, and irrefutable importance. And, its importance for women is especially significant as they seek to achieve independence, from both men and the state: education provides them with means to a career, to an escape from patriarchal structures both within and outside the home, to economic, emotional, and familial well-being, and to decision-making over their lives. Education is "the prerequisite for improvement in women's status" (Tinker 1990, p. 33), and the implications of its worth extend way beyond the immediacy of one's participation in it.

Given the significance of higher education, we draw on the work of Amartya Sen and place postsecondary education within the capability approach, a set of ideas about the basic quality of life standards, bringing attention to its central importance in securing freedom, enhancing national and individual well-being, and developing capabilities in individuals by increasing the options available to them (Sen 1999). Sen's intent in the capability approach is, in part, a rejection of traditional approaches that invoke the claim that a unified, singular focus on increasing income, usually by increasing out-of-home work, will resolve issues facing people who are poor. His claims rest on a larger, more macro notion of what poverty means in one's daily life—of what it means to one's sense of self, to one's health and well-being, to one's children and family, to one's community, and to one's nation. In essence, what it means to *not* have the opportunity to be what one wants to be. Sen's concerns are grounded in a deep understanding that social and institutional arrangements limit, and in many cases, restrict people's, and importantly women's, ability to realize a life of value. The most fundamental democratic and human right, Sen would argue, is the ability of the individual to *invent* herself, free from the constraints of poverty and authoritarianism.

To this end, education—higher education—is key. Education is not only essential to freedom but is a major source of women's empowerment. It not only shapes her destiny but also enables her to help others in a meaningful way, both by maximizing individual potential and by rectifying greater injustices including the indoctrination, through systems of education, of the subordinate status of women in society. It is a public good, an investment in the future of a society, facilitating, perhaps even ensuring, women's economic participation and hence her freedom and independence (Sen 1999).

Our focus in this chapter is on low-income women with children; our intent is to embrace the fundamental notions of the capability approach in arguing that postsecondary education must be afforded all citizens, especially poor women with families because of what it enables one "to do or to be" (Sen 1982, 1985, 1992, 1993, 1999). We offer insights into how the lives of poor women in the United States have been altered by their participation in postsecondary educational programs. Guided by notions encompassed by the capability approach we explore these womens' sense of increased intrinsic worth—empowerment, self-esteem, well-being; the remarkable impact higher education has had on their lives—employment opportunities, new friends, ideas and ambitions, expanded life choices; the implications of changed relationships with their children as they model aspirations for enriched and fulfilling lives; and their commitment to affiliation—enhanced motivations to contribute to and participate in their communities and the society at large (Dreze and Sen 1995).

Higher Education and Well-Being: Promoting Human Capabilities

Opening doors and realizing potential are what higher education should be all about. And yet, with the passage of the Personal Responsibility and Work Opportunity Reconciliation Act (PRWORA) in 1996 by the U.S. Congress, access to a postsecondary education was severely restricted for people on welfare. Welfare reform—as PRWORA is commonly called—dramatically changed "welfare as we knew it." "Work first" ideology, the bedrock of the federal welfare legislation, pushed the merits of postsecondary education for low-income women to the background, despite the well-known, interlocking relationship between higher education and women's earnings, employment, and well-being.

Immediately prior to the act's passage, however, advocacy groups throughout the United States cautioned the Congress on the punitive nature of this new legislation, challenging traditional, stereotypical, individualistic claims about welfare receipt by placing it in context with the broader societal problems of gender inequity, unemployment, low-wage jobs, and poverty. Everywhere the same message was reiterated: *welfare reform must be about raising families out of poverty, and this won't happen until we change the economic position of women in the labor market.* Securing access to postsecondary education for low-income families was high on the list. In Maine, a determined and systematic effort to change the terms of the debate by insisting on the crucial connections among women's economic security, labor market access, income-generating capacity, success, and postsecondary education created a political climate receptive to legislation that honored these connections. As a result, in 1997, the legislature created the Parents as Scholars (PaS) program, a state-funded student aid program for welfare recipients, making Maine one of only two states (Wyoming is the other) in the nation at the time to include access to postsecondary education in the state welfare plan. PaS, developed in response to federal dictates, within the restrictions on access to higher education imposed by the federal law, formed the basis of a coherent policy that not only maintained an obligation to women's education but could also more successfully accomplish PRWORA's avowed claim to "end dependency" and secure "self-sufficiency."

In an extensive literature review focused on women, welfare, and higher education, Washington D.C.'s Center for Women Policy Studies analyst Erika Kates reported that "low income women who have engaged in higher education experience several tangible advantages: their incomes improve, their level of satisfaction with their own lives and their children's improves; they become more productive citizens; and they become prime motivators in improving the lives of others closely connected to them." (Kates 1992, p. 2) Our own discussions with the Maine Department of Human Services' (DHS) staff and PaS program participants revealed similar findings: participants had increased self-esteem and confidence, fewer family crises, and experienced positive family interactions around issues related to education. Children of participants experienced a heightened quality of life and had elevated aspirations for and comfort with higher education. The DHS staff found that participants required fewer support services and less employee time and energy. Employers have access to a better-educated and well-rounded workforce. The State of Maine sees the genuine prospect of higher earning power and a stronger tax base as well as a viable citizenry.

Higher Education Matters

There is little doubt that higher education can increase earning capacity and help women trying to escape poverty (see Blau 1998; Gittell, Gross, and Holdaway 1991a, 1991b, 1993; Gittell et al. 1989; Kates 1996; Nettles 1991; Welfare Reform Network News 1998; Thompson 1993). Education is especially important in enabling women and people of color to increase their earning capacity, stemming the tide of an increasingly segregated workforce. And higher education is crucial for families that are poor. Without it, low-wage work, known to be highly vulnerable to high worker turnover rates as well as high rates of unemployment and underemployment, exacerbates already-desperate family situations. Furthermore, low-skilled women leaving public assistance for employment often cycle back onto welfare when their hours are reduced or they are laid off; women are less likely to return to welfare if they have college degrees (Gittell, Gross, and Holdaway 1993; Kates 1996).

In an exhaustive contemporary study tracing trends in the well-being of North American women from 1970 to 1995, economist Francine Blau (1998) affirmed positive associations between educational attainment and labor force participation, increased earnings, and general well-being. While women made substantial progress during those 25 years, it was the rising rate of women's participation in higher education that made the most difference. Findings revealed real wage gains for female college graduates (20.3 percent) compared to women with high school degrees or some college (8–9 percent), and declines for high school dropouts (2.2 percent). Rising educational attainment was also a factor in women's increasing labor force participation: rates increased 19 percent among college-educated women and 29 percent among the most highly educated women, while among the least-educated women labor force participation rates rose by only 4 percent (Blau 1998, pp. 124–125, 131).

The potential that higher education affords an individual over a lifetime is tremendous; quality education provides one the *ability* to do things in the future (Sen 1999). For women, the opportunity afforded them to escape oppressive homes, seek outside employment, and secure their own and their families' financial future is often immeasurable. Sen's work stresses that economic self-sufficiency, equality, and gender equity for women cannot occur without reasonable access to education and literacy in general and, by inference, higher education in particular. His view, however, stands in stark contrast to the fundamental tenets of U.S. welfare policy, which promotes a disinvestment in human capital and discourages personal achievement by dissuading, and in many cases prohibiting, poor women from seeking access to higher education.

Higher Education—Welfare Policy

When Aid to Dependent Children, the precursor to Temporary Assistance for Needy Families (TANF) program and Aid to Families with Dependent Children (AFDC), was first established within the Social Security Act of 1935, women raising children alone were provided financial benefits to enable them to remain home to care for their children. The provision of care for children in these single-parent—most of whom were widowed—mostly white, families was of central concern to the act's architects. No programs were established for workplace training or advanced education: women were thought to belong in the home to care for their children. Traditional notions of women as caretakers and nurturers, not as providers or workers, dictated these times.

Over time, the population of what became known as "welfare recipients" grew and its ethnic and racial composition and marital status changed, some would say quite dramatically. Simultaneously, welfare policy grew more stringent, restrictive, and prescriptive. The initial aim of keeping women in their homes to care for their children gave way to requirements forcing them to work outside the home, handing over the care of their children to others. Over this same period, yet for very different reasons, higher education became receptive to and encouraged applications from women. Decades of research and scores of studies now document the undeniable, positive impact of women's education on earnings, success, achievement, and individual and national well-being.

Yet in 1996 when President Bill Clinton signed the PRWORA, access to higher education for low-income parents, mostly women, on welfare was rescinded. It made two further, significant changes to the welfare system: (1) it established a 5-year lifetime limit on benefits, thereby eliminating a 61-year-old entitlement to cash assistance for low-income mothers and children and creating TANF, and (2) it required welfare recipients to work in exchange for benefits. Welfare recipients, among the poorest and most vulnerable people in the country, were now confronted with an almost-unthinkable and daunting challenge: "end dependency" and "become self-sufficient" without access to postsecondary education.

The principal intent of PRWORA, to move mostly poor women off "welfare" and into jobs, was promulgated "to end the dependency of needy parents on government benefits by promoting job preparation . . . and work to enable them to leave the program and become self-sufficient" (Personal Responsibility and Work Opportunity Reconciliation Act 1996). Federal law discouraged States from allowing recipients to meet federal work participation requirements by attending college. Predictably, these restrictions had a devastating impact on the three-quarters of a million welfare recipients enrolled in postsecondary colleges: decreases in enrollment among recipients throughout the country ranged from 29 percent to 82 percent (Finney 1998, p. 2). Federal restrictions and corresponding sanctions forced many college-bound women to leave school for work, despite the long-term consequences. The fear of federal financial reprisal, coupled with the political hazards inherent in the failure to follow the path of tough, work-based reform, led most states to abandon postsecondary education programs for welfare recipients. And while welfare reform supporters were quick to point out that the new law did not prohibit welfare recipients from attending college, the powerful financial consequences for states were evident in their implementation plans. Even in those states where the law theoretically allowed welfare recipients to go to college, welfare caseworkers were under such intense pressure from state administrators to meet work participation rates that many ignored the law and required recipients to work in addition to going to school (Schmidt 1998; see also Greenberg, Strawn, and Plimpton 2000,1999).

Nationally over four million families were impacted, most headed by women. While the legislation afforded states tremendous leeway in determining how they would handle their welfare caseloads and amidst a plethora of research studies confirming the relationship between higher education and increased earnings potential, 48 states opted to restrict recipients' access to higher education. At present, we know from numerous studies of the consequences for welfare recipients who were denied access to higher education: minimum- and low-wage work in unstable markets; high rates of return to welfare; less ability to meet basic family needs— food, shelter, clothing, heat, light, medicine; increased homelessness among families and among children; poorer families at greater risk.

The PaS Program: An Enactment Overview

In 1997, the Maine legislature rejected the route prescribed by PRWORA and boldly chose to institutionalize access to postsecondary education for low-income parents. As one of only two states to recognize the necessity of including higher education in its welfare plan, it enacted the PaS program after hearing testimony from many welfare recipients about the value of education and about their difficulties in obtaining it. Maine's approach presumed that when PaS families leave welfare they will earn higher wages, be more likely to have employment-based health insurance, be less likely to return to welfare, have greater family stability, and provide better modeling and educational aspirations for their children.

PaS is a separate, state-funded student aid program limited to 2,000 TANF-eligible participants enrolled in two or four year postsecondary education programs. They receive the TANF-equivalent cash benefits and supportive services through Additional Support for People in Retraining and Employment (ASPIRE), Maine's work-to-welfare program: assistance with child care, transportation, car repairs, auto liability insurance, eye care, dental care, books and supplies, clothing and uniforms, occupational expenses, and other services as necessary. The participants are also eligible for services from the postsecondary institutions they attend: personal counseling, on-campus health care, job opportunities, job search assistance, campus housing, child care, financial aid, support groups, academic advising, and wellness programs. Because PaS is state-funded, the federally mandated five-year time limit does not apply. Students are required to "work" but can count both school and study time toward these hours: Maine broadened the definition of work to include school-related volunteer activities, tutoring, internships, and work-study, all count toward the required work hours. Students are expected to complete their programs in one and one-half times the normal matriculation time and maintain at least a 2.0 grade point average (GPA).

PaS: Nearing the Capability Approach

The significance and benefits of higher education opportunities for poor women cannot be overstated. Findings from surveys of PaS program participants we have conducted affirms the positive correlations between access to higher education, well-being, empowerment, earnings, and enhanced relationships with children (see also Butler and Deprez 2002; Deprez and Butler 2001; Deprez, Butler, and Smith 2004; Polakow et al. 2004). Documenting the impact of PaS on the lives of participants and providing data to federal and state legislators and policymakers about the real circumstances of impoverished families in the state has been central to our efforts to help preserve program funding in the face of state and federal budget deficits and competing demands for diminishing revenue. Additionally, we wanted to inform the federal debate on welfare reauthorization by providing a case study of the positive consequences of providing welfare participants with access to postsecondary education. We collected two consecutive waves of data from participants in the program, providing a preliminary view of the impact of postsecondary education on the lives of one group of women on welfare.

In 1999 we began collecting data on the experiences and lives of PaS participants when we sent a 19-page survey to each of the 848 adults enrolled in the program. The instrument comprised questions about the participants' current and past educational experience; work and welfare receipt history; health and the health of their children; current financial situation; children and child care circumstances; time use in their daily lives; experience in the program and with their postsecondary educational institution; and beliefs about how PaS had impacted their lives. We received 222 completed surveys—a 26.2 percent response. The findings from this rich data set reflected the positive impact of higher education on the lives of respondents: participants wrote about improved self-concepts, greater opportunities, and enriched family relationships, or put another way, valued capabilities and functionings.

We were also interested in following these respondents over time to determine if the initial positive impact of higher education translated into increased potential for economic self-sufficiency—something we knew would be of key importance in both state and federal policy debates. In June 2000, we contacted the 192 original respondents to ask if they would be willing to remain in our study and provide us with information that would help us to stay in contact with them: nearly two-thirds of this sub-sample (66.1 percent) were located and replied affirmatively to our inquiry. Seventeen months later—in November 2001—we sent a second survey to these 127 individuals and received 65 completed surveys, giving us a response rate of 51.2 percent, testimony to participants' eagerness to report on their successes and their interest in participating in the state discussion on the future of the PaS program.

This second survey was similar to the first, asking questions about employment, education, child care, health, and well-being; exploring how PaS had impacted respondents' feelings about themselves, relationships with family members, and future plans. It was also intended to assess how PaS graduates were faring in the labor market: these data would be particularly important in the federal TANF reauthorization debates and in state legislative discussions about the future of the PaS program.

Profile of PaS Participants

In 1999 all but seven of the respondents were women (96.8 percent), and most were either single (36.9 percent) or divorced (34.7 percent). They ranged in age from 20 to 56, with a median age of 30. All had biological children living with them: most had just one child (43.2 percent) or two children (34.2 percent). Forty-five respondents (20.3 percent) reported that they had a disability and fifty-six (25.2 percent) said they had a child with a disability. In the three years prior to entering the PaS program, the vast majority of respondents had been employed in either full-time or part-time work (81.4 percent); the median hourly wage was $6.50. Most respondents (83.6 percent) had received traditional high school diplomas before entering postsecondary education; the remainder had completed their high school studies in alternative, out-of-school-based programs and received a General Equivalency Diploma (GED). Respondents were fairly evenly split between pursuing two- and four-year degrees, though slightly more were in four-year programs (56 percent). Respondent GPAs ranged from 1.83 to 4.0—five respondents reported 4.0 averages—with a mean GPA of 3.21. Respondents seemed happy with their postsecondary education institutions with 92.5 percent rating their institutions as either good or excellent.

In the 2001 survey, all 65 respondents were women. They ranged in age from 23 to 44, with a median age of 31. Less than half the sample (44.6 percent) reported being partnered, married, or remarried, while the majority stated they were either single (32.3 percent), or separated or divorced (23 percent). All the respondents were caring for children, and nearly half (49.2 percent) had preschool children. The overwhelming majority (80.0 percent) reported that they were the sole carers of their children. In this survey, we expected that some respondents would no longer be students or PaS participants, and this was in fact the case. Less than one quarter of the respondents (23.1 percent) was still in school and still PaS participants. Another five respondents (7.7 percent) reported being in school, but no longer with the program, and four respondents (6.2 percent) had left both school and the program. Well over half of the respondents had graduated (63.0 percent) and all but six of these individuals were employed at the time of the survey. While respondents had majored in a broad range of subjects, over half of the respondents were concentrated in three predominantly female fields: social services (24.6 percent), health professions (15.4 percent), and business (e.g., accounting; 15.4 percent). The majority was or had been in four-year degree programs (60 percent), the remainder in two-year programs. The respondents' cumulative GPA ranged from 2.5 to 4.0 with a median GPA of 3.4. Most PaS students juggled full course loads, parenting, and part-time employment while maintaining these high grades.

The data we collected provide new evidence of the importance of education for greatly expanding the opportunities for and enriching the lives of recipients of welfare. The narratives support past research that show receiving a post-secondary degree can be life changing on many levels for low-income parents: they experience increased self-esteem, they become role models for their children; they gain skills that open up employment opportunities; and they are motivated to contribute to society. Considerable data was collected in these surveys on a number of topics, among them wages, family composition, health, and housing. We report in this chapter on the narrative responses to selected open-ended questions included in the two surveys as they reflect the importance of the premise of the capability approach: "to enable poor women to function fully in society and family" (Harcourt 2001, p. 4).

Change in Self-Concept

The first of several open-ended questions asked, "Has your participation in a postsecondary program affected your feelings about yourself?" Nearly all the respondents answered this question affirmatively, including narrative comments describing changes in their self-concept. Three dominant themes emerged in the responses—"empowerment," "self-esteem," and "well-being"—reflective of the overwhelmingly positive, transformative experience of being in school. Citing initial nervousness and insecurity as they began classes, nearly all reported that these melted away as they met the challenges of higher education and often exceeded their own expectations.

Many respondents (57.5 percent) reported a feeling of independence and liberation as a result of their participation in postsecondary education. For example, some women spoke specifically of how they would no longer allow abusive men to push them around and tell them what to do. A 30-year-old divorced mother of three children, a third year sociology student with the career goal of being a teacher, formerly in an abusive relationship, spoke of her increased feeling of empowerment:

> I feel confident in my ability to face challenges. I have ambition. Before I used to crumble in the face of adversity and wait for someone to "rescue" me—either my parents or a boy friend. Now I know I can make it on my own. I don't have to "settle" anymore.

Over 55 percent of respondents reported increased self-esteem, greater confidence, and strengthened self-respect as a result of going back to school. One 27-year-old woman working on her bachelor's degree in human services, caring for a 7—year-old child who suffered from migraines and was diagnosed with Attention Deficit Hyperactivity Disorder, wrote:

> I have gained so much confidence in myself through this experience. Before I began school I felt absolutely worthless, as if I was a total waste of life. I am now on the Dean's list and have been from semester one! I am very proud of my efforts and achievements, I cannot imagine how life would be if I had not entered this institution.

Another respondent, a 39-year-old senior majoring in political science, said:

> My self-esteem has greatly improved. For most of my life I believed I was not intelligent enough to go to college. When I began school I was very nervous and stressed about whether I could succeed; I have! I now feel confident in my ability to think, process and produce answers both academically and personally.

This mother of two teenagers maintained a 3.7 GPA and hoped to go on to law school.

The final theme that emerged indicated a heightened sense of well-being; 76 respondents (38 percent) mentioned feeling happier and healthier and found school to be fun. The following quote from a 34-year-old divorced woman with three children reflects this sense of well-being. This woman had nearly completed her degree in plant and soil technology and looked forward to more job stability than she had had in the past, when she had had to hold down several jobs at a time.

> I feel much more self-confident and comfortable. I have new friends and a social life. I feel like a complete person now that my life has direction and goals. I am much happier and my children are also. I feel smarter and very encouraged about the future. I have goals and hope now. I have gotten to understand myself more deeply and am becoming a much better person.

Closely related to the change in self-concept brought about by higher education was the change in anticipated life prospects for the respondents.

Impact on Lives

A second question asked respondents whether postsecondary education had changed their lives or enhanced their opportunities. Almost all (99.5 percent) of the respondents felt college had changed their lives. The three themes that emerged were "opportunity," "goals," and "enrichment." Narratives reflected a mood of anticipation and excitement. The respondents believed their hard work in school would pay off in terms of opening doors to employment that would be personally satisfying and financially lucrative, and that would bring them respect. Being in school had introduced them to new ideas, new friends, new ambitions, and new ways of looking at the world. Seventy-five percent wrote about the increased opportunities that their postsecondary training afforded them by opening doors to better jobs, increased pay, and more stable employment. This theme is illuminated by the words of a senior nursing student, a 25-year-old mother of two children with severe health problems—one with epilepsy and the other with asthma—who maintained a 3.4 GPA:

> Without the education I have received I would still be cocktail waitressing for $2.13 per hour and tips until 3:30 am, with a BSN I will have the opportunity to provide for my children, work for good pay, receive employee benefits, and have a marketable career, especially since I chose the field of nursing.

Almost 50 percent of the respondents wrote about how college was helping them both meet their current goals and create new goals for the future. Some respondents stated that they were realizing their dreams, while others spoke of the dreams that college was helping them establish. A junior, majoring in human development, said:

> I realize the potential that I have had within me and feel that I can and will attain all of the goals I have set for myself. I have so much to offer my future employers and clients.

This student parent with three special needs children—two with mental health diagnoses and one with a behavioral impairment—planned to make her career working with children with behavioral problems.

On a final theme, nearly a quarter of the respondees (23.1 percent) indicated that postsecondary education had changed their lives, broadened their horizons, enriched their lives, and allowed them to see things in new ways. This theme of enrichment is illustrated by the words of a senior in behavioral sciences, a mother of four children, aged between six and ten, who hoped to pursue a master's degree in occupational therapy upon finishing her undergraduate degree.

> I became a senator on student government. I learned about politics when once I hadn't a clue. I've learned so much about so many different things. I'm really excited about the spectrum of opportunities open to me and the

easy availability of learning opportunities to do those things I want to. Life sure is good these days.

Not surprisingly, the changes in self-concept and life goals impacted the relationships respondents had with their children, partners, parents, and friends. As one element of the system shifts, other elements must adjust.

Effect on Relationships

One of the final survey questions asked respondents whether postsecondary education had affected their relationships with their children: a majority reported that it had. Two primary and two lesser themes emerged: "role model" and "understanding", and "pride" and "stress." This set of themes was less homogeneous in the sample than the two previous sets, illustrating the complexity of being a single parent while going to school full-time. While some very positive effects were reported regarding the impact of going to school on parent/child relationships, many respondents wrote of the difficulty of having too little time to be with their children. This multilayered phenomenon is described below.

About one third of the respondents (33 percent) spoke about how their return to school modeled good decisions and behavior for their children. Some spoke of doing homework together with their children, others said their children were doing better in school, and yet others said their children now spoke of going to college themselves. A 37—year-old student of business management said her return to school had impacted her two children, aged nine and ten:

> It has inspired them to get a good education and has shown them that they can aspire to be what they want to be career wise. It has shown them that to be self-sufficient they must work towards their careers and that education is a life-long journey.

This respondent, who had a 3.85 GPA, had never worked before returning to school—she claimed she never had enough education; now she expected to own her own business after completing her degree.

Similarly, over one quarter of the respondents (26.2 percent) wrote about the better understanding they had of their children, and the higher quality time they were spending together. They had learned more about child development, enhanced their parenting skills, and gained patience. They spoke of enjoying their children more, talking with them more, and having more fun with them. A small but significant number of respondents (16 percent) mentioned the pride their children expressed about their pursuit of higher education. An additional 31 respondents spoke of their parents' and partners' pride. Many reported being the first person, for some the first woman, in their families to attend college. This statement by a junior, who is a mother of three children, exemplifies this feeling of pride by her family.

> My children are proud of me. They did learn to go to college after high school though. They said they don't want to wait because it is harder when you get older. They have seen me struggle and I think they respect that. They are incredible kids. My dad is extremely proud. I'll be the first female of both sides to ever graduate from college.

While the overwhelming number of respondents reported positive associations with going to college, 13 percent reported that their relationships had been negatively affected. They wrote of being overly stressed, not having enough time for their children or other family members. Some respondents worried about their children growing up without them, others knew this stress was time-limited and for a good purpose. The theme of "stress" was significantly related to geographic location: rural respondents were more likely to write about the stress in their lives than their urban counterparts. Pursuing higher education while raising children, often alone, was evidently rewarding and beneficial to family relationships for many respondents. The stress of demanding academic work combined with parenting responsibilities is not an unexpected finding and merely strengthens the case for providing adequate supports such as child care and transportation assistance for these families as they work toward economic security.

In the 2001 follow-up survey, respondents were again asked how participation in a postsecondary education program had affected their feelings about themselves, their lives, their opportunities, or their relationships with their family members. Over two-thirds of respondents reported unequivocally and enthusiastically about the positive effects of their education (69.6 percent). The remainder (30.4 percent) indicated more ambivalence about the experience and their futures. Going to school can be very stressful, especially while raising children alone with few financial resources. Although most respondents clearly stated that the benefits outweighed the hardships, several said that they wondered at times if the final outcome would be worth all the sacrifices they made along the way. In contrast to ambivalence about their postsecondary experience, however, nearly half of all the respondents (49.2 percent) took the time to write lengthy narrative responses about the positive benefits of education in their lives. The themes in these narratives were similar to those that had emerged from the qualitative data collected in the first PaS survey two years earlier (Deprez, Butler, and Smith 2004; Butler and Deprez 2002; Deprez and Butler 2001).

Enthusiasm emerged as a key theme among the majority of respondents. Returning to school was repeatedly reported as positively affecting self-esteem, employment opportunities, and family relationships. A 29-year-old graduate with an associate's degree in business writes about how her life has turned around; she accomplished things that she had previously never thought she was capable of. Her narrative exudes enthusiasm about her new life:

> I never thought I was smart enough to go to college . . . what distorted thinking that was! I realized that I was actually quicker at getting things down than I thought. . . . All I needed was one small success after another to realize my worth. I've become a better mother, a better partner, a better friend. I've gained a lot of confidence in that I can really make a difference in people's lives because of the knowledge I've gained.

Another respondent, in an abusive relationship as she started school, similarly reports on the dramatic change in her life during her years in college. At the time of the survey, this woman worked part-time as a school-based educator earning $10.00 per hour; she had earned her bachelor's degree in social sciences a year earlier. Along with her new partner, she was raising four preschool children, two of whom had long-term disabilities.

> When I started college, I was in an abusive relationship . . . I had no self-esteem, but wanted desperately to make life better. . . . He'd sabotage the car and offer to baby-sit but then not be around. I prayed and prayed for him to die or go back to jail. I couldn't make him go away no matter what I said or did. Finally my prayers were answered. He went to jail. I had tremendous difficulties focusing on my studies and critical thinking. . . . I figured I was an educated woman; this stuff shouldn't happen to me. I kept plugging away and was amazed when I got good grades . . . My life has changed significantly because of school. I used to have trouble in social situations. . . . It was hell. I was also very angry and didn't like being a mom. Now, I'm a school-based educator with my local rape crisis center. I go to school to teach. I would never have imagined myself doing this. I love my family now and do my best to provide for their emotional needs. I'm a Girl Scout leader for a third year. The town I live in calls me whenever there's a social function, to help out . . . If you had known me when I first started school, you'd understand just how much life has changed for me as well as those who are with me daily.

Two divorced respondents in their late 30s, graduates, employed at the time of the survey, reflected on how their return to school had opened doors for them in terms of work, and provided good role modeling for their children. The first, a graduate with a four-year degree in legal studies, was working two jobs: a full-time position in a juvenile drug court, for which she made over $2,000 per month, and an on-call position bringing her an additional $200 each month. She described her experience as follows:

> Returning to college was one of the best decisions I ever made. It certainly built my self-esteem. . . . I had no idea that I could do so well. . . . I met new friends and learned how to parent better. I was able to make contacts that led to several job offers during college and after. I am proud of the role model I have been able to become for my daughter. By earning my bachelor's degree, I have become self-sufficient.

The second respondent had completed her associate's degree in business with honors; she secured a secretarial job with benefits, earning nearly $19,000 per year. Prior to returning to school, she had never been employed. She characterized her college experience as transformative:

> My decision to return to school or continue my education affected many areas of my life. . . . My days were restructured. My skills were refined . . . My self-confidence and self-esteem grew well beyond what I could have ever imagined. There was an adjustment period during my reentry to school and my family life changed in many ways. Most of this was positive. My relationship with my children changed also. Their mother was now a student in addition to all the other roles I played in their lives. I can only hope and pray that they saw me as a positive role model. At one point, both of them said they wanted to go to college "just like mama." . . . After graduation, many doors were opened to me. . . . I am grateful for the opportunity to continue my education and I am still pursuing going on to get my bachelor's degree in my field. Thank you for this opportunity. I am confident that I will be successful in whatever happens in my life even though it requires hard work, determination and perseverance.

Another theme evident in these narratives, which was also evident in the 1999 survey responses, was that of giving back to the community. One respondent above reported being a scout leader, called upon by the town when they needed help with social functions. The desire to give back to society led another respondent to continue her education beyond graduation in order to get state certification as a teacher. Her narrative captures the essence of the themes threaded through so many of the narratives: improved self-esteem, a sense of accomplishment, richer employment opportunities, role modeling for children, wanting to make a contribution to society, and a sense of security about the future:

> Completing my degree has had a tremendous impact on my self (image, self-esteem, etc.) and my future. When I received my degree this past spring I finally felt that I had achieved something worthwhile in life. . . . At the same

time I want to be able to give back to my community. I decided to go into teaching and am working on getting state certified . . . My family is proud and supportive and my child has a positive role model for the effort one must make when working toward a goal. I will always be grateful to have had the opportunity to complete my education. That is the one factor that can keep me from sliding into hopeless poverty. In completing my degree I knew that I can do whatever it takes to go after my dreams.

These data provide new evidence of the importance of education for greatly expanding the opportunities for and enriching the lives of recipients of welfare. The data confirm earlier research that a postsecondary degree is life changing on many levels for low-income parents: they experience enhanced self-esteem, become role models for their children, gain skills that open up employment opportunities, and are motivated to contribute to society.

Diana Spatz, executive director of LIFETimE—an advocacy organization in California devoted to helping low-income women access education—said that higher education allows us to talk about "intergenerational promise" as opposed to "intergenerational dependence," the phrase so often used by policymakers (Spatz 2000). These findings speak to that promise. Women in the PaS program report feeling empowered to reach new goals. They have greater self-respect, broader horizons, feel healthier and happier, relate better with their children, and, in the words of one respondent, "education has become a family value" for them. Their participation in the program has enabled them to both see and experience a more expansive range of available life choices. It is this expansion of choice, which Sen so rightly promotes, that extends one's own freedom and secures the ability to model that freedom for others.

Prospects for the Future

Higher education is inextricably linked to quality of life; it can make a substantial difference in the lives of poor women *and* address deep-rooted causes of poverty. It plays a role, Sen argues, "not only in accumulating human capital but also in broadening human capability" (Saito 2003, p. 24; Dreze and Sen 1995, p. 43). Yet, social welfare and economic policies in the United States have consistently failed to alleviate poverty among women, most particularly among those heading households. In part this has occurred because of institutionalized assumptions that "reforms are gender neutral" (Aslanbeigui, Pressman, and Summerfield 1994, p. 6). The high personal and social cost for women who are expected to fulfill family roles of both breadwinner and nurturer is often unacknowledged; the work they have been forced to do affords neither them nor their families security. Findings from a 1998 study, which examined the relationship between family structure and children's educational and occupational success over four decades, affirmed that the "negative effect of single mother families on children's socioeconomic attainment is more a function of mothers' disadvantaged *occupational position* and less a function of their disadvantaged *employment status*' (Biblarz and Rafferty 1999, p. 344 [italics ours]). Another recent, complementary study revealed that 69 percent of all welfare recipients have skills that qualify them for some postsecondary education, which would enable them to increase their advantage in the labor market, position them for job advancements, and secure their family's stability and security (Carnevale and Desrochers 1999, p. 7).

That higher education was discouraged in PRWORA is in marked contrast to current societal attitudes about education: nearly nine out of ten Americans agree that college education has become as important as a high school diploma once was (Morin 2000). If welfare "reform" and the presumable and subsequent well-being of poor women is to be measured by reduction in poverty rates and by sustained and permanent movement into the primary labor force, immediate attention needs to be focused on areas intimately tied to the success of this venture. Access to higher education ranks at the top of the list, although access alone is still not enough to change gendered asymmetries in both private and public spheres. Issues such as assuring pay equity, raising the minimum wage, eliminating job segregation, increasing union affiliation, promoting labor market opportunities—especially in rural areas—stabilizing benefits, and securing availability of supportive services also demand our immediate and collective attention.

Maine's PaS program provides welfare recipients with access to postsecondary education programs that can increase their prospects of a life without poverty, and the opportunity to develop valued capabilities to choose the life they wish to lead. While the program is not an absolute guarantee of a life without poverty, low-income women and their families face greatly enhanced chances of secure living when their opportunities for obtaining and maintaining successful, supportive, and fulfilling work are increased. Education affords women the best possibility for meaningful participation in a democratic society, enabling them to overcome injustice, to enhance the well-being of the nation. It reveals and demonstrates ways in which capabilities expand and inform each other—its multidimensional impact—within the capability approach. It has the potential to address what Sen refers to as "the more inclusive idea of capability deprivation" (1999, p. 20); and it has the ability to replace long-standing and "typical patterns of employment and education for women," as one generation transmits its experiences to another (Sen 1990, p. 137). Support for this concept comes from a study by David Cotter and associates, substantiating the earlier findings of Rae Lesser Blumberg that found "the continued expansion of women workers with college degrees may be the only prospect of undoing gender as a major force in the organization of work" as it is "the demand for women's labor in productive work that is important in diminishing gender inequality" (1998, p. 1675; see also Blumberg 1978, 1984).

The difficulty of operationalizing the capability approach in the United States, apart from securing support from elected officials on its singular importance, is that the ability to secure higher education is, for the most part, out of the hands of poor women. The competing demands on their time, attention, and already limited resources along with the soaring costs of tuition reduce the prospect of higher education as an option in their lives if it means foregoing basic necessities—food, shelter, clothing, heat, medicine—for them and their families. Hence, their ability to participate in postsecondary education programs must be coupled with societal support and secured resources for these necessities if higher education is to secure a place in their decision-making schemas. To not provide them this choice undermines the integrity of this nation, one that purports to believe in individual responsibility and the merits of an educated citizenry, and relegates poor women and their children to perpetual patterns of lives lived in poverty.

Institutions of higher education, welfare activists, feminist academics and scholars, policy institutes, think tanks, and advocacy organizations must move to ensure access to postsecondary education for some of our most vulnerable citizens. Otherwise "the avowed objective of welfare reform . . . to lift welfare recipients out of poverty by moving them into paid employment," reveals only dismal prospects for the future. We must assume our responsibility to work on behalf of those left out of educational systems to secure their entry, "to act in some way that makes people's lives better" ("Educating for the Lives of Others" 2000, p. 7). In Maine, the PaS program has modeled one way in which the long quest to reduce poverty, by fostering individual human capabilities among women from poor families can be achieved, affording low-income women who are parents, the same access to higher education that the public at large believes is essential for this society to succeed in the future.

References

Aslanbeigui, Nahid, Steve Pressman, and Gale Summerfield. 1994. *Women in the age of economic transformation.* London: Routledge.

Biblarz, Timothy J., and Adrian E. Rafferty. 1999. Family structure, educational attainment, and socioeconomic success: Rethinking the pathology of matriarchy. *American Journal of Sociology* 105:321–365.

Blau, Francine D. 1998. Trends in the well-being of American women, 1970–1995. *Journal of American Economic Literature* 36:112–165.

Blumberg, Rae Lesser. 1978. *Stratification: socioeconomic and sexual inequality.* Dubuque, IA: William C. Brown.

———. 1984. A general theory of gender stratification. In *Sociological theory,* edited by R. Collins. San Francisco: Jossey-Bass.

Butler, Sandra, and Luisa S. Deprez. 2002. Something worth fighting for: Higher education for women on welfare. *Affilia: Journal of Women and Social Work* 7:30–54.

Carnevale, Anthony P., and Donna M. Desrochers. 1999. *Getting down to business: Matching welfare recipients' skills to jobs that train, Executive Summary.* Princeton, NJ: Educational Testing Services.

Cotter, David A., JoAnn Defiore, Joan Hermsen, Brenda M. Kowalewski, and Reeve Vanneman. 1998. The demand for female labor. *American Journal of Sociology* 103:1673–1712.

Deprez, Luisa S, and Sandra Butler. 2001. In defense of women's economic security: Securing access to higher education under welfare reform. *Social Politics* 8:210–227.

Deprez, Luisa S., Sandra Butler, and Rebekah J. Smith. 2004. Securing higher education for women on welfare in Maine. In *Shut Out: Low income mothers and higher education in post-welfare America,* edited by V. Polakow, S. S. Butler, L. S. Deprez and P. Kahn. New York: SUNY Press.

Dreze, Jean, and Amartya Sen. 1995. Basic education as a political issue. *Journal of Educational Planning and Administration* 9:1–26.

"Educating for the lives of others: An interview with Martha Nussbaum. 2000. *On Campus with Women, AAC&C* 30 (1):1–16.

Finney, Johanna. 1998. Welfare reform and post-secondary education: research and policy update. *IWPR Welfare Reform Network News* 2 (1):1–9.

Gittell, Marilyn, Jill Gross, and Jennifer Holdaway. 1991a. Education in a democratic society: from the 1960s to the 1980s. In *Women, work and school: Occupational segregation and the role of education,* edited by L. Wolfe. Boulder, CO: Westview Press.

———. 1991b. Women on welfare: education and work. In *Women, work and school: Occupational segregation and the role of education,* edited by L. Wolfe. Boulder, CO: Westview Press.

———. 1993. *Building Human Capital: The impact of postsecondary education on AFDC recipients in five states. Report to the Ford Foundation.* New York: Howard Samuels State Management and Policy Center, the City University of New York Graduate School.

Gittell, Marilyn, Jill Gross, Jennifer Holdaway, and Janice Moore. 1989. Denying independence: Barriers to the education of women on AFDC. In *Job training for women: The promise and limits of public policies,* edited by S. Harlan and R. Steinberg. Philadelphia, PA: Temple University Press.

Greenberg, Mark, Julie Strawn, and Lisa Plimpton. 1999. *State opportunities to provide access to postsecondary education under TANF.* Washington, D.C.: Center for Law and Social Policy.

———. 2000. *State opportunities to provide access to postsecondary education under TANF.* Washington, D.C.: Center for Law and Social Policy.

Harcourt, Wendy 2001. The capabilities approach for poor women: Empowerment strategies towards gender equality, health and well-being. Paper read at the Conference on Justice and Poverty, Cambridge, England.

Kates, Erika. 1992. *Women, welfare and higher education: A selected bibliography.* Washington, D.C.: Center for Women Policy Studies.

———. 1996. Colleges help women in poverty. In *For crying out loud: Women's poverty in the United States,* edited by D. Dujon and A. Withorn. Boston, MA: South End Press.

Morin, Richard. 2000. What Americans think. *The Washington Post Weekly Edition,* May 8, 34.

Nettles, Saundra Murray 1991. Higher education as the route to self-sufficiency for low-income women and women on welfare. In *Women, work and school: Occupational segregation and the role of education,* edited by L. R. Wolfe. Boulder, CO: Westview Press.

Personal Responsibility and Work Opportunity Reconciliation Act. 1996. Pub. Law No. 104-193, *U.S. Statute at Large* 110 Stat. 2105 codified at 42 U.S.C. sec. 601 1998: 42 U.S.C. sec. 601a2 and 602a1Ai1998.

Polakow, Valerie, Sandra S. Butler, Luisa S. Deprez, and Peggy Kahn, eds. 2004. *Shut out: Low income mothers and higher education in post-welfare America.* New York: SUNY Press.

Saito, Madoka. 2003. Amartya Sen's capability approach to education: A critical exploration. *Journal of Philosophy of Education* 37 (1): 17–33.

Schmidt, Peter 1998. States discourage welfare recipients from pursuing a higher education. *Chronicle of Higher Education* 44 (20): A34.

Sen, Amartya. 1982. *Choice, welfare and measurement.* Oxford: Blackwell.

———. 1985. *Commodities and capabilities.* Amsterdam: New Holland.

———. 1990. Gender and cooperative conflicts. In *Persistent inequalities,* edited by I. Tinker. New York: Oxford University Press.

———. 1992. *Inequality re-examined.* Oxford: Oxford University Press.

———. 1993. Capability and well being. In *The quality of life,* edited by M. Nussbaum and A. Sen. Oxford: Oxford University Press.

———. 1999. *Development as freedom.* Oxford: Oxford University Press.

Spatz, Diana. 2000. Personal communication with authors, June 26.

Thompson, Joanne J. 1993. Women, welfare and college: The impact of higher education on economic well-being. *Affilia: Journal of Women and Social Work* 8:425–441.

Tinker, Irene. 1990. The making of a field In *Persistent Inequalities,* edited by I. Tinker. New York: Oxford University Press.

Welfare Reform Network News. 1998. Post secondary education leads to increased earnings for welfare recipients. *Newsletter of the Institute of Women's Policy Research.*

Part III

Taking the Capability Approach Forward in Education

12

Conclusion: Capabilities, Social Justice, and Education

Elaine Unterhalter and Melanie Walker

The chapters in this collection have considered how the capability approach helps us to think about some key issues in education and schooling. While there are similarities with the perspective on education associated with the capability approach and other views of education linked to Aristotelian notions of human flourishing, what is new about the capability approach is that it combines a normative idea with consideration of how to link this with practice not just in education but in a wide range of political, economic, and social fields that bear on education. Thus, for example, utilizing the capability approach in education illuminates thinking about questions of justice and the distribution of schooling, gender equality, redressing poverty, politics, the link between school and the labor market, policymaking, education measurement, institution building, management, and pedagogies. It has been used to address specific questions in schools, the problems of regional and national policymaking, and issues about global relations. It combines well-articulated philosophical underpinnings with a versatile range of applications. However, as many of the chapters conclude, work on the capability approach in education is just beginning and a very wide range of issues are unexamined. There are also a number of areas where the capability approach is problematic and critical discussion is still required. In concluding this book we want to highlight how some of these might be taken forward by researchers and practitioners in education and those working more generally on the capability approach.

The challenges posed in education and the solutions devised are often context specific, but there are a number of features of contemporary discussion in education that recur across different countries and groups. The capability approach seems to offer a useful conceptual vocabulary for some, but not all, of these settings. The political economy of globalization locates education as a key commodity, a particular attribute of people incorporated in global social relations, an area of national contestation and competition, a platform from which to develop better health, employ-ability, higher skill levels, better levels of social understanding. Social jus-tice reflections on education talk to issues of distribution and the relationship between education and society and contested versions of "the good life," in particular concerns with global relations, the citizenship of states, engagement with civil society, and intimate relations between friends and within families. Education is heavily imbued with aspiration, deeply contested with regard to content and organization, and spectacu-larly under-resourced for those who have least. The social justice scholar-ship on education mixes a particular blend of optimism and critique. The capability approach combines these two features, which is one aspect of its explanatory reach, but it is silent on a number of issues of significance. In reviewing some emerging problems in research on the capability approach and education we want to point out how much the framework talks to contemporary concerns, and also address selected problems associated with unsettled issues and silences.

Capabilities and Rights

What are the differences between capabilities and rights? Is the technical language associated with the capability approach a step backward from the language of rights that has widespread popular currency and conveys many of the same aspirations? The concept of rights, like capabilities, expresses ethical concerns for each and every individual. Rights, like capa-bilities can be viewed as requiring both negative and positive freedoms for their realization. Rights are sometimes schematically grouped to distin-guish rights to education, rights in education, and rights from education. One could make a similar classification of capabilities.

The capability approach we think, both supports a human rights dis-course but also goes beyond it in demanding that we ensure not only rights, be these conceptualized legally or morally, but also people's capa-bilities and functionings. Thus, not only is the right to equal opportunities for students in education important, but also the capability to function as participants in equal-opportunity educational processes and outcomes. For example, women in the UK have won the right to higher education, but we need to ask about the capabilities of these women to access, partic-ipate, and succeed in universities, in other words, to ask both about their right to education and their capability to be educated and to lead a life they have reason to value. Formal rights need to be implemented and secured in practice, and the language of capabilities, because it is different from rights, alerts us to this.

Nussbaum (2000) distinguishes rights from capabilities, suggesting that the institutional locus of rights is too constraining for the broader conception of human flourishing both within and beyond institutions that she develops for capabilities. Some rights, she argues, entail combined capabilities, and she generally emphasizes that rights are constitutional commitments whereas capabilities require attention to implementation and evaluation of such rights (Nussbaum 2000).

Amartya Sen develops a philosophical underpinning for the notion of rights, as ethical demands entailing wide imperfect obligations (Sen 2004). He also suggests a complex relationship between rights and capabilities.

Human rights are best seen as rights to certain specific freedoms, and the correlate obligation to consider the associated duties must also be centered on what others can do to safeguard and expand these freedoms. Since capabilities can be seen, broadly as freedoms of particular kinds, this would seem to establish a basic connection between the two categories of ideas (Sen 2005, p. 152). The argument Sen develops entails an analytic separation between two aspects of rights: substantive opportunities that are best understood as capabilities and process freedoms that are intrinsic to a notion of rights and a theory of justice, but does not play the same central role in conceptualizing capabilities (Sen 2005, p. 156). Thus, for example, in assessing the young women with different capability sets described in chapter 1, there are differences in capabilities, but not in functionings between them. But questions remain about what process freedoms each had. Say, we know that the girl from Nairobi decided to prioritize her leisure activities with friends, although she had ample chance to participate in discussions in a number of different locations about the links between gaining a pass grade in mathematics and securing access to courses she wanted to study at university; understanding issues about her own health such as risk factors associated with different forms of contraception, the effects of climate change in her country, and the ways in which her government reported on its work to combat poverty. Sen would argue that she had not suffered a diminution of capabilities or rights understood as process freedoms, but had she not been offered opportunities to engage in these processes of discussion and deliberation, and were efforts not made to secure against poverty, or the worst effects of climate change, or the risks associated with certain forms of contraception, or obstacles to women gaining entry to higher education, there would be limitations on human rights. Thus rights are features of persons, but they are also aspects of social arrangements whether or not this particular rights bearer is currently in a position to benefit from them.

Harry Brighouse and Ingrid Robeyns have conceptualized the link between rights and capabilities somewhat differently. Robeyns suggests that rights should be viewed as one possible instrument to reach a goal of expanding capabilities. In some contexts rights might be the most appropriate instrument, but in others public discussion or the portrayal of the issues through film or theater might do this task better (Robeyns 2005, pp. 82–83). Brighouse sees capabilities as the basis of rights claims (Brighouse 2004, p. 82). If a parent, for example, claims that she has the fundamental right to secure only a particular form of schooling for her children linked with her interpretation of the tenets of her faith, a question in terms of justice could be posed as to whether that form of right was required to serve the capability to be educated. If the capability to be educated could be secured through a different kind of schooling there would be no violation of fundamental rights, although the parent might argue that other capabilities, for example, to bring up a child as an accepted member of a religious community were being constrained. There is no space in this conclusion to settle this particular debate. Here we want to signal that the relationship between rights and capabilities is a matter of keen discussion. (For further work on this theme, but without drawing on education examples, see Osmani 2005; Vizard 2006.)

It is evident that the discussion of the links between rights and capabilities has particular resonance for that stream in education studies currently concerned with questions of citizenship, religious freedom, and toleration; the relationship between state and civil society in education provision; and the interpretation of the Universal Declaration of Human Rights that protects parents' rights to choose the education of their children. The problem of trade-offs between different capabilities discussed in a number of chapters may be amenable to elucidation if rights and capabilities are distinguished in particular settings. However, it is also clear that specifying more precisely the relationship between capabilities and rights in education may not have particular relevance for discussions of education and economic growth, and may not be adroit enough with questions of cultural diversity and the range of responses to this.

Global Justice, National Competition, and the Question of Institutions

Both Sen and Nussbaum have been engaged in writing on the theme of global justice and education (Sen 2004; Nussbaum 2003b, 2005). In this we can locate them with an emerging and distinctive group of academic advocates of cosmopolitanism (Held 2004; Beck 2006; Appiah 2006; Brock and Brighouse 2005). Indeed the capability approach is sometimes considered as providing some of the appropriate normative framing for cosmopolitanism and the attempt to further develop global institutions to address poverty or the denial of education (Unterhalter 2007, forthcoming; Vizard 2006). However, the relative silence in writing on the universalism of the capability approach to consider particular ties of group affiliation, citizenship, or national identity have been criticized (Stewart 2005; Mandle 2006). One of the key debates in writings on cosmopolitanism concerns the significance of obligations to citizens of one's own state, citizens of other countries, or those classed as peoples without the protection of citizenship. Sen and Nussbaum both refuse to draw distinctions between people based on citizenship, but many of their critics do.

In its articulation of thick cosmopolitanism, the approach refuses to draw distinctions on the basis of national boundaries. This may be a considerable shortcoming in thinking about education where national identity has generally been bound up with the form of the education system. Indeed a number of writers note how the capability approach fails to pay sufficient attention to institutional arrangements that secure capabilities, be these international, national, or local, and have advocated consideration of how work on justice and institutions complement Sen's capability approach (Deneulin, Nebel, and Sagovsky 2006; Robeyns 2004; Richardson 2006).

This debate concerning capabilities, global obligations, and the form of institutions seems particularly generative for thinking about the way multilateral organizations are involved with education in multiple settings, debates about increases in aid to fund education provision in the poorest countries, and the nature of higher education under conditions of globalization where there are high levels of mobility among elite teachers and students. It also appears that the capability approach is useful in helping to sort out arguments that demand countries build education systems only to advance national skills development and competition. But the capability approach probably has less to say about the particular design of forms of knowledge transfer or the implications of the commodification of education.

What Kind of Education? Lists, Capabilities, and a Mild Perfectionism

The question of the content of education the nature of school knowledge and what comprises the school curriculum has been of central concern to those working in education for decades (Peters 1966; White 2005; Hirst 1974;

Young 1998; Bernstein 2001). The nature of the national curriculum is a topic of heated contemporary debate in most societies. Does the capability approach throw any new light on an old debate?

One critique of the approach is that there are some sharp divisions between key writers on whether or not we can list essential capabilities and therefore use that list as a guide to thinking about curriculum. How problematic for work in education is the standoff in the literature on the capability approach on lists and indexing? As we outlined in chapter 1, there is considerable distance between Sen's open-endedness and refusal to specify a "canonical" list of capabilities and Nussbaum's development of a list of universal central functioning capabilities. Nonetheless, despite Sen's commitment to the importance of democratic deliberation in selecting capabilities, he has articulated a particular vision of how education in the UK needs to enhance our capacity to live together in times marred by violence and suspicion across lines of race and religion:

> Education is not just about getting children, even very young ones, immersed in an old inherited ethos. It is also about helping children to develop the ability to reason about new decisions any grown-up person will have to take. The important goal [in thinking about the introduction of faith schools] is not some formulaic "parity" in relation to old Brits with their old faith schools but what would best enhance the capability of the children to live "examined lives" as they grow up in an integrated country.
>
> (Sen 2006, p. 160)

We can see that Sen, here in the guise of a public intellectual at a moment marked by sharply contrasting views on faith schools, is arguing for a particular content and form of schooling, not just a relativist position that any kind of education agreed by a family or a community will do. He argues that faith schools constrain reasoned identity choices and their agency because "young children are placed in the domain of singular affiliations well before they have the ability to reason about different systems of identification that may compete for their attention" (2006, p. 9).

We agree and consider that the capability approach entails a kind of mild perfectionism in thinking about education. This is much less prescriptive than Nussbaum's list, but nonetheless we consider the approach entails recognizing plurality of and in education, and also means that not everything claiming to be educative counts as education or effective and life-enhancing learning. We cannot leave the matter of capabilities in the case of education entirely open for children and young people because education affects their continuing journey into and through adult life and having a good life. It matters therefore what it is that students are learning, and what they are learning to be in school. It then follows that the opportunities we open up or foreclose in education for students' capability development matters. Thus, content and sequence cannot be seen as endlessly plural and open for negotiation.

Considering people as subjects—agents—of their own lives is central in the capability approach. Sen (1999, p. 18) emphasizes "the ability of people to help themselves and to influence the world." A lack of agency or a constrained agency equates to disadvantage. It is thus key in education that we promote freedom and agency to participate further in education and social debate and to enlarge wider freedoms. What capabilities support agency development is then a matter of public deliberation, including in books such as this one.

For example, Brighouse (2000, p. 65) argues that autonomy should be a fundamental value in the design of educational policy; and "all children should have realistic opportunity to become autonomous adults," because autonomy "enhances dramatically the ability of individuals to identify [for themselves] and live lives that are worth living" (p. 88). Social justice, he says, "requires that each individual have significant opportunities to live a life which is good" (p. 68). It then also follows that children need to develop a sense of what it means to live well, to be able to compare different ways of life, and to choose a good life for themselves. This in turn involves fostering the capability for critical reflection on one's own goals and values as "an essential part of living well" (p. 67). Children should, argues Brighouse, learn how to assesstruth, weigh up evidence, investigate, and think about their decisions, and so learn a "critical attention" to the options available to them. However, Brighouse is careful to distinguish between an autonomy-facilitating and an autonomy-promoting education:

> The argument claims that equipping people with the skills needed to reflect on alternative choices about how to live is a crucial component of providing them with substantive freedom and real opportunities, by enabling them to make better rather than worse choices about how to live their lives. The [autonomy-facilitating] education does not try to ensure students employ autonomy in their lives, any more than Latin classes are aimed at ensuring that students employ Latin in their lives. Rather it enables them to live autonomously should they wish to.
>
> (2000, p. 80)

Brighouse's idea of an autonomy-facilitating education is similar to Nussbaum's (2000, p. 78) human capability of "practical reason," which she describes as "being able to form a conception of the good and to engage in critical reflection about the planning of one's life." She insists that in the interests of democracy and tolerance in society, children "not be held hostage to a single conception [of the good life]" (2003a, p. 42), but that they are exposed to other ways of living. Her concern is that they should have the capability to critically reflect and plan, and if they so choose to opt for a nonautonomous life in which this capability will not be exercised, for example, if a life in a traditional religious community is chosen. She argues that the state has no business telling adults "that they are not leading worthwhile lives." Thus, adults might choose to live nonautonomously. Nussbaum is therefore clear "that we shoot for capabilities, and those alone [i.e., not for functionings]. Citizens must be left free to determine their own course after that" (2000, p. 87). For Nussbaum this includes choosing a nonautonomous life. In other words, she does not claim that personal autonomy is the only valuable style of life. But for her, a life lacking the *capability* of practical reason "is not a life in accordance with human dignity" (2003a, p. 39); but this is different from insisting that we should all function in this way.

Power, Curriculum, and Pedagogy

The question of power and its implication both for discussions of curriculum and pedagogy has also not yet received wide enough coverage in the scholarship on the capability approach and education, although some of the chapters in this book begin to address this (and see Walker 2006). We suggest that consideration of power as a form of critique and in particular the examination of power in relation to curriculum and pedagogies needs to be integrated with capabilities for equality and justice. Adding in other theories is not incompatible with the approach, which does not lay claim to being a theory of social justice or equality, but rather a framework requiring additional theories for specific contexts, such as applications in education (Robeyns 2003). In the end, Sen is not an educational theorist, although we believe that his ideas have much to offer educational analysis and thinking well about the problems of and in education. Nonetheless, Sen arguably underestimates the force of history, power, and conflict in learning arrangements. He certainly underestimates that education in formal settings may not be an unqualified good for all students (Unterhalter 2003). Nussbaum (2003b) claims transformative power for education in women's lives, but does not fully consider how education is to bear the weight of such transformation. What is missing in both Sen's and Nussbaum's writing on education is the sense of history and struggle in the formation of learner identities in pedagogical spaces in the face of dominant education norms and values and learning practices permeated by power, history, language, and contradiction (Contu and Willmott 2003). If learning "involves the construction of identities" (Lave and Wenger 1991, p. 53), it is also suffused with power and practices of exclusion and inclusion.

Thus we need also to draw on and integrate sociological concepts, for example, theorizations of power and participation in educational settings; of learner identity formation as suffused with history, power, and conflict; and of critical thinking and criticality in which learners develop an awareness of discursive practices and power relations in order to exert more conscious control over their everyday lives.

To take just one example. We might "add-in" to analysis drawing from the capability approach ideas from critical pedagogy, which broadly draws on critical theory and a structuralist critique of schooling as a site for the reproduction of capitalism and capitalist values and social relations, but also views schooling as a space of resistance and change. Such ideas help us to understand that schooling is a major site of cultural practice and the recognition or devaluing of personal and social identities, values, and abilities (Baker et al. 2004). All education is a site of symbolic control, where "consciousness, dispositions and desire are shaped and distributed through norms of communication which relay and legitimate a distribution of power and cultural categories" (Bernstein 2001, p. 23). The pedagogical is then fundamentally about the production of learner identities, "or the way one learns to see oneself in relation to the world" (Grande 2000, p. 471). This formation of self is "one of the core struggle concepts of critical pedagogy" (Ibid.) and through this concern with the formation of self we then take up learning in relation to structures of race, class, gender, disability, and so on. Bourdieu (Bourdieu and Passeron 1977) has pointed out how higher education tacitly requires students to work with and through middle-class language codes and socially constituted dispositions, which they are assumed to already possess, and which are not made explicit or taught in a systematic way in education. In this way education reproduces inequalities and privilege through the means of a "racism of intelligence" (Bourdieu 1993, p. 177).

Critical pedagogy offers the theoretical tools to examine power and domination in education and uncover its selection and sorting functions. It critiques not just educational hurdles to educational success and agency, but also how these barriers are shaped by social, economic, and political obstacles to social justice and democracy. It posits a theorization of identity as fluid, contingent, dynamic, and in-process so that change is always possible and new identities might be produced at the nexus of contradictions and struggle through learning for a critical subjectivity. It requires us to consider "the interplay between power, difference, opportunity and institutional structure" (Grande 2000, p. 49).

Acknowledging contestations over power as a key theme in the consideration of education raises a number of sharp issues for the capability approach. Far from capability development being smooth and seamless, it might demand justified anger" (Nussbaum 2000, p. 79) in speaking back to authority and a history that works to keep some students in their place. Far from social relations in learning being easy and predictable, coming to know the other is difficult, fraught, and while desirable for learning together, ultimately also a nearly impossible project. Where there are inequalities structured into learning situations, it is also *not* the case that the marginalized can rely on the privileged to identify with them in a process of mutual capability building. In such cases the freedoms of some are not then also the freedoms of others. Becoming an agent through learning is not straightforward. As Bernstein (2001) and Lave and Wenger (1991) have argued, learning processes are suffused with the exercise of power and control, and knowledge is mediated and acquired through processes of participation and social interaction, in schools and universities, in which not all are accorded equal power, equal recognition, and equal esteem.

Critical pedagogy is a reminder that we need to be mindful of power in public dialogues. Under current education arrangements, all students do not have fair opportunity to access, participate in, and progress through education. It is not the case that individual rationality is sufficient to overcome a conflict of interpretations and identities. It is arguably not the case that the capability approach really addresses the space of power in which capability might form and develop from positive learning in education. Young (2000) importantly reminds us that deliberative democratic processes are a form of practical reasoning and hence, one might suggest, fundamental to capability development as freedom. But she also points out that we need explicit attention to connectedness and the inclusion of the dependent and vulnerable to enable collective problem-solving by all those significantly involved in or affected by a decision, and under conditions of dialogue that allow diverse perspectives and opinions to be voiced.

Critical pedagogy with its concern for the oppressive effects of education and of the power dimension of pedagogy therefore offers a significant resource. Like critical pedagogy, the capability approach locates education within a larger social and human development vision. Both share a concern for the potential of education to enable the doing of good in the world. Both approaches embrace complexity. Both have concern with the voices "of those who have to struggle to be heard" (Kincheloe 2004, p. 24). Critical pedagogy is better at showing how power works and that education may be oppressive as well as transformative. Critical pedagogy is sharper at dealing with contextual dynamics of language, discourse, and power. Critical pedagogy has a stronger conceptualization of collective as well as individual agency in learning so that "individual criticality is intimately linked to [increasing] social criticality" (Burbules and Berk 1999, p. 55). At the same time, the concern in the capability approach with what we can actually do and be grounds critical pedagogy in processes of learning and equality of learning outcomes and the connection between learning, education, and other processes of social change.

Measuring Education

Some of Sen's earliest work in putting the capability approach to work in relation to problems of global social justice related to issues of measurement. Prompted by Mahbub ul Haq, former finance minister of Pakistan, working for the UN in the late 1980s, Sen began to think about an approach to measurement of quality of life or capabilities as an alternative to the measures of commodities that were standard in economics (Sen 2003, pp. viii–ix). While a wide range of actors within the UN system were interested in this problem and discussion of quality of life and capabilities was well advanced (Nussbaum and Sen 1993), Sen was worried that the project of distilling a one-number index of quality of life was a deformation of the richness of the aims of the project.

> On one occasion when I told Mahbub that the one-number index I was then devising at his request could not but be quite "vulgar" because of the inescapable oversimplification that he was demanding, Mahbub replied: "we need a measure of the same level of vulgarity as the GNP, but one that is not as blind to social aspects of human life as the GNP is."
>
> (Sen 2003, p. xiii)

Sen then did develop the one-number index, the Human Development Index (HDI), and the United Nations Development Programme (UNDP), from 1990, has published annual rankings of countries in relation to their HDI as well as other elements of comparison, most notably the Gender Development Index (GDI), and the Gender Empowerment Measure (GEM), and the Human Poverty Index (HPI). The annual publication of the *Human Development Report* is widely reported in the press all over the world and countries watch their ranking with keen interest.

Ul Haq (1995) outlined the rationale for the HDI taking the form that it did, as follows. The index was to measure the basic concept of human development (which became the popular term for capabilities) and was to include only a limited number of variables in a composite index measuring social and economic choices. The HDI comprises a measure of life expectancy at birth; a measure of a country's achievement in literacy and the combined enrolment rates at primary, secondary, and tertiary education; and GDP per capita. Each component index is weighted equally (UNDP 2005, p. 341). Later refinements (the GDI) measured gender in terms of women's and men's relative position in life, education, and earnings, while the GEM was a composite measure comprising women's political participation, economic participation, and power over economic resources. The HPI looked at the probability of a child surviving to 40, adult literacy rates, the percentage of the population with a decent standard of living (in terms of access to water and food), and the extent of social exclusion or long term unemployment (UNDP 2005).

It can be seen that the education measures in the HDI have the same connotations of "vulgarity" that Sen suggested to ul Haq affected the whole project of the search for a simple number to measure human flourishing. Working in education it is evident that enrolment is not a useful measure of whether people learn and that the many different interpretations of literacy make this a particularly imprecise indicator of knowledge and the capacity to engage in practical reason. Nonetheless, the authors of the *World Development Reports* have continued to use these measures because these are the numbers that are routinely collected by governments and multilateral agencies.

Partly in imitation of the *Human Development Reports*, UNESCO began to publish *Global Monitoring Reports* from 2002 reviewing progress on Education for All (EFA) (UNESCO 2002). These have come to include detailed statistics on enrolment, teacher qualifications, progression through different levels of school, and aid flows. A version of the HDI in education, the Education for All Development Index (EDI) was developed (UNESCO 2003, p. 284).

Thus the achievements of the capability approach in generating new approaches to measurement and monitoring are considerable. But there are a wide range of debates. Do the existing measures of education used by UNDP and UNESCO adequately capture a notion of education, as opposed to presence at school, as argued in some of the papers in part 1? Should alternative measures be generated from the bottom up or through some process of considering patterns of valued learning opportunities across societies? (Young 2006). Is the trade off between the rich information needed to adequately assess capabilities and the actual information collected inadmissible, as Roemer (1996) has suggested, arguing that "good enough vulgarity" is far too distorting. Are the manipulations of existing data to better capture, say, gender equality in education (Unterhalter, Challender, and Rajagopalan 2005; Unterhalter 2006) a help or a hindrance to thinking about measurement in education linked to the capability approach?

While people interested in the capability approach tend to discuss these issues among themselves, the general work of measuring and monitoring education performance proceeds without reference to the thinking linked to the capability approach, relying on standard analyses of the education production function, or multilevel models of the value-added components of schooling. There is thus scope for considerable development in bringing these two streams of discussion of measurement into closer dialogue with each other.

To Conclude

Finally, it is important to acknowledge the genuinely radical ideas for education in the capability approach—not only its concern with hetero-geneity and actual living out of valued lives, but also its call for *both* redistribution of resources and opportunities *and* recognition and equal valuing of diversity along intersecting axes of gender, social class, race, ethnicity, disability, age, and so on. It thus integrates distributional, recognitional, and process elements of justice. It argues for each and every person having the prospect of a good life, that they have reason to value, by enabling each person to make genuine choices among alternatives of similar worth, and to be able to act on those choices. Moreover, a particular strength in the capability approach is that, while broadly oriented to justice, through its emphasis on capability (potential to function) it does not prescribe one version of the good life but allows for plurality in choosing lives we have reason to value. The approach emphasizes the importance of capability over functioning—not a single idea of human flourishing, but a range of possibilities and a concern with facilitating valuable choices. Above all, the capability approach offers a freedoms-focused and equality-oriented approach to practicing and evaluating education and social justice in all education sectors and in diverse social contexts. While there remains much exciting work to be done and gaps, such as those we have outlined here to be taken up, we hope that this book will be a stimulus to thinking well about the capability approach, social justice, and education.

References

Appiah, K. A. 2006. *Cosmopolitanism: Ethics in a world of strangers.* New York: Norton.

Baker, J., Kathleen Lynch, S. Cantillon, and J. Walsh. 2004. *Equality: From theory to action.* Houndmills: Palgrave Macmillan.

Beck, U. 2006. The *cosmopolitan vision.* Cambridge: Polity.

Bernstein, B. 2001. Symbolic control: Issues of empirical description of agencies and agents. *International Journal of Social Research Methodology* 4 (1): 21–33.

Bourdieu, Pierre. 1993. *Sociology in question.* London: Sage.

Bourdieu, Pierre, and J.-C. Passeron. 1977. *Reproduction in education, society and culture.* 2nd ed. London: Sage.

Brighouse, Harry. 2000. *School choice and social justice.* Oxford: Oxford University Press.

————. 2004. *Justice.* Cambridge: Polity.

Brock, G., and Harry Brighouse. 2005. *The political philosophy of cosmopolitanism.* Cambridge: Cambridge University Press.

Burbules, N., and R. Berk. 1999. Critical thinking and critical pedagogy: Relations, differences, limits. In *Critical Theories in Education,* edited by T. S. Popkewitz and L. Fendler. New York: Routledge.

Contu, A., and H. Willmott. 2003. Re-embedding situatedness: The importance of power relations in learning theory. *Organization Science* 14 (3): 283–296.

Deneulin, Severine, M. Nebel, and N. Sagovsky, eds. 2006. *Transforming unjust structures: The capability approach.* Dordrecht: Springer.

Grande, S. 2000. American Indian geographies of identity and power. *Harvard Educational Review* 70 (4): 467–498.

Held, D. 2004. *Global covenant.* Cambridge: Polity.

Hirst, P. H. 1974. *Knowledge and the curriculum.* London: Routledge and Kegan Paul.

Kincheloe, J. L. 2004. *Critical pedagogy.* New York: Peter Lang.

Lave, J., and E. Wenger. 1991. *Situated learning: Legitimate peripheral participation.* New York: Cambridge University Press.

Mandle, Jon. 2006. *Global justice.* Cambridge: Polity.

Nussbaum, Martha C. 2000. *Women and human development: The capabilities approach.* Cambridge: Cambridge University Press.

————. 2003a. Political liberalism and respect: A response to Linda Barclay. *Nordic Journal of Philosophy* 4:25–44.

————. 2003b. Women's education: A global challenge. *Signs: Journal of Women and Culture in Society* 29 (2): 325–355.

————. 2005. *Frontiers of justice: Disability, nationality, species membership.* Cambridge, MA: Belknap Press of Harvard University Press.

Nussbaum, Martha C., and Amartya Sen, eds. 1993. *The quality of life: Studies in development economics.* Oxford: Oxford University Press.

Osmani, S. R. 2005. Poverty and human rights: Building on the capability approach. *Journal of Human Development* 6 (2): 205–220.

Peters, Richard Stanley. 1966. *Ethics and education.* London: Allen & Unwin.

Richardson, Henry. 2006. Trust, respect, and the constitution: The social background of capabilities for freedoms. Paper read at the International Conference of the Human Development and Capability Association: Freedom and Justice. August 29–September 1, Groningen, the Netherlands.

Robeyns, Ingrid. 2004. Justice as fairness and the capability approach. Paper read at the Fourth International Conference on the Capability Approach: Enhancing Human Security. September 5–7, Pavia, Italy.

————. 2005. The capability approach: A theoretical survey. *Journal of Human Development* 6 (1): 93–114.

Robeyns, Ingrid 2003. Sen's capability approach and gender inequality: Selecting relevant capabilities. *Feminist Economics* 9 (2–3): 61–91.

Roemer, John. 1996. *Theories of distributive justice.* Cambridge: Harvard University Press.

Sen, Amartya. ————. 1999. *Development as Freedom.* Oxford: Oxford University Press.

———. 2003. Foreword. In *Readings in human development*, edited by S. Fukuda-Parr and A. K. S. Kumar. New Delhi: Oxford.

———. 2004. Elements of a theory of human rights. *Philosophy and Public Affairs* 32 (4): 315–356.

———. 2005. Human rights and capabilities. *Journal of Human Development* 6 (2): 151–166.

———. 2006. What clash of civilizations? Why religious identity isn't destiny. *Slate Magazine*. http://www.slate.com/id/2138731/(accessed September 18, 2006).

Stewart, F. 2005. Groups and capabilities. *Journal of Human Development* 6 (2): 185–204.

ul Haq, Mahbub. 1995. *Reflections on human development.* Oxford: Oxford University Press.

UNDP. 2005. *Human development report 2005: International cooperation at a crossroads—Aid, trade and security in an unequal world.* New York: UNDP. http://hdr.undp.org/reports/global/2005/pdf/HDR05_complete.pdf (accessed September 7, 2005).

UNESCO. 2002. *EFA global monitoring report, 2002: Is the world on track?* Paris: UNESCO. http://www.unesco.org/education/efa/monitoring/monitoring_2002.shtml.

———. 2003. *EFA global monitoring report. Gender and education for all: The leap to equality.* Paris: UNESCO. http://portal.unesco.org/education/en/ev.php-URL_ID=23023&URL_DO=DO_TOPIC&URL_SECTION=201.html.

Unterhalter, Elaine. 2003. The capabilities approach and gendered education: An examination of South African complexities. *Theory and Research in Education* 1 (1): 7–22.

———. 2006. Global inequalities in girls and women's education: How can we measure progress? Paper read at the International Conference of the Human Development and Capability Association: Freedom and Justice, August 29–September 1, at Groningen, the Netherlands.

———. Forthcoming 2007. *Gender, schooling and global social justice.* Abingdon/New York: Routledge.

Unterhalter, Elaine, Chloe Challender, and Rajee Rajagopalan. 2005. Measuring gender equality in education. In *Beyond Access: Developing gender equality in education*, edited by S. Aikman and E. Unterhalter. Oxford: Oxfam.

Vizard, P. 2006. *Poverty and human rights.* Oxford: Oxford University Press.

Walker, Melanie. 2006. *Higher education pedagogies: A capabilities approach* Maidenhead: SRHE/Open University Press and McGraw-Hill.

White, J. 2005. *The curriculum and the child.* Abingdon/New York: Routledge.

Young, Iris Marion. 2000. *Inclusion and democracy.* Oxford: Oxford University Press.

———. 2006. Defining valued learning and capability. Paper read at the International Conference of the Human Development and Capability Association: Freedom and Justice, August 29–September 1, Groningen, the Netherlands.

Young, Michael F. D. 1998. *The curriculum of the future.* London: Falmer Press.

Bibliography

General briefings on the capability approach and much work in progress can be found on the website for the Human Development and Capability Association (HDCA): http://www.capabilityapproach.com/ Home.php.

Further details of the thematic group working on education and the capability approach (coordinated by Lorella Terzi and Elaine Unterhalter) can be found at: http://www.capabilityapproach.com/Thematic.php?sid=cbffbe1b29207d706d e8450bfa2d3dd8&grpcode=thematic2.

For annotations on many of the texts listed below, please consult either the HDCA site or http://k1.ioe.ac.uk/schools/efps/elaine/Capability-and-Education.pdf.

Books

Aikman, Sheila, and Elaine Unterhalter, eds. (2005) *Beyond Access: Transforming Policy and Practice for Gender Equality in Education.* Oxford: Oxfam Publications.

Flores-Crespo, Pedro (2005) Educación superior y desarrollo humano. El caso de tres universidades tecnológicas. Mexico: Asociación Nacional de Universidades e Instituciones de Educación Superior.

Gold, Anne (2005) *Values in Leadership.* Issues in Practice. London: Institute of Education .

Walker, Melanie (2006) *Higher Education Pedagogies: A Capabilities Approach.* Maidenhead: SRHE/Open University Press and McGraw-Hill.

Walker, Melanie, and Elaine Unterhalter, eds. (Forthcoming 2007) *Amartya Sen's Capability Approach and Social Justice in Education.* New York: Palgrave.

Chapters in Books

Patel, Ila (2003) "Literacy as Freedom for Women in India." In *Literacy as Freedom: A UNESCO Roundtable,* edited by N. Aksornkool. Paris: Literacy and Non-formal Education Section, Division of Basic Education.

Raynor, Janet (Forthcoming 2007) "Schooling Girls: Easing Burdens in Rural Bangladesh?" In *Gender, Education and Development: Contemporary Frameworks, Methodologies and Policy Agendas,* edited by S. Fennel and M. Arnot. Oxford: Routledge.

Unterhalter, Elaine (2005) "Fragmented Frameworks? Researching Women, Gender, Education and Development." In *Beyond Access: Transforming Policy and Practice for Gender Equality in Education,* edited by S. Aikman and E. Unterhalter. Oxford: Oxfam.

——— (In press) "The Capability Approach and Gendered Education: Some Issues of Operationalisation in the Context of the HIV/AIDS Epidemic in South Africa." In *Operationalising Amartya Sen's Capabilities Approach,* edited by S. Alkire, F. Comim, and M. Quizilbash. Cambridge: Cambridge University Press.

——— (Forthcoming 2007) "The Political Foundations of Gender Equality in Education, Needs, Rights and Capabilities." In *Gender, Education and Development: Contemporary Frameworks, Methodologies and Policy Agendas,* edited by S. Fennel and M. Arnot. Oxford: Routledge.

Unterhalter, Elaine, Chloe Challender, and Rajee Rajagopalan, eds. (2005) "Measuring Gender Equality in Education." In *Beyond Access: Transforming Policy and Practice for Gender Equality in Education,* edited by S. Aikman and E. Unterhalter. Oxford: Oxfam.

Walker, Melanie (2004) "Pedagogies of Beginning." In *Reclaiming Universities from a Runaway World,* edited by M. Walker and J. Nixon. Maidenhead: SRHE/Open University Press.

——— (Forthcoming 2007) "Widening Participation and Lifelong Learning." In *Philosophical Perspectives on Lifelong Learning,* edited by D. Aspin. Dordrecht: Springer Press.

Watts, Michael (2006) "What *Is* Wrong with Widening Participation in Higher Education." In *Discourse, Resistance and Identity Formation,* edited by L. Roberts, W. Martin, and J. Satterthwaite. Stoke on Trent: Trentham Books.

Watts, Michael, and David Bridges (2006) "Enhancing Students' Capabilities? UK Higher Education and the Widening Participation Agenda." In *The Capability Approach: Transforming Unjust Structures,* edited by S. Deneulin, M. Nebel, and N. Sagovsky. Dordrecht: Springer Verlag.

Young, Marion (Forthcoming 2007) "Capability as an Approach to Evaluation of Learning Outcome from Local Perspectives." Milan: Fondazione Giangiacomo Feltrinelli.

Journal Articles

Arends-Kuenning, Mary, and Sajeda Amin (2001) "Women's Capabilities and the Right to Education in Bangladesh." *International Journal of Politics, Culture, and Society* 15, no. 1: 125–142.

Enslin, Penny, and Mary Tjiattas (2004) "Liberal Feminism, Cultural Diversity and Comparative Education." *Comparative Education* 40, no. 4: 503–516.

Flores-Crespo, Pedro (2002) "En busca de nuevas explicaciones sobre la relación entre la educación y la desigualdad. El caso de la Universidad Tecnológica de Nezahualcóyotl." *Revista Mexicana de Investigación Educativa* 7 no. 16 (September–December): 537–576. This article has been translated into English

in the same journal, see: http://www.comie.org.mx/revista/Abstracts/carpeta%
2016/16labsTem4.htm.

——— (Forthcoming) "Education, Employment and Human Development: Illustrations from Mexico." *Journal of Education and Work.*

——— (Forthcoming) "Ethnicity, Identity and Educational Achievement: Lessons Drawn from Mexico." *International Journal for Education and Development.* Special issue edited by David Johnson and Frances Stewart.

Jackson, William A. (2005) "Capabilities, Culture and Social Structures." *Review of Social Economy* 63, no. 1: 101–124.

Klasen, Stephan. (2001) "Social Exclusion, Children and Education. Implications of a Rights-Based Approach." *European Societies* 3, no. 4: 413–445.

Nussbaum, Martha (2006) "Education and Democratic Citizenship: Capabilities and Quality in Education." *Journal of Human Development* 7, no. 3: 385–398.

——— (2004a) "Liberal Education and Global Community." *Liberal Education,* Winter.

——— (2004b) "Women's Education: A Global Challenge." *Signs: Journal of Women and Culture* 29:325–355.

Robeyns, Ingrid (2006) 'Three Models of Education: Rights, Capabilities and Human Capital.' *Theory and Research in Education* 4, no. 1: 69–84.

Saito, Madoka (2003) "Amartya Sen's Capability Approach to Education: A Critical Exploration." *Journal of Philosophy of Education* 37, no. 1: 17–33.

Terzi, Lorella (2005) "Beyond the Dilemma of Difference: The Capability Approach on Disability and Special Educational Needs." *Journal of Philosophy of Education* 39, no. 3: 443–459.

——— (2005) "A Capability Perspective on Impairment, Disability and Special Needs: Towards Social Justice in Education." *Theory and Research in Education* 3, no. 2: 197–223.

Unterhalter, Elaine (2003a) "The Capabilities Approach and Gendered Education: An Examination of South African Complexities." *Theory and Research in Education* 1, no. 1: 7–22.

——— (2003b) "Crossing Disciplinary Boundaries: The Potential of Sen's Capability Approach for Sociologists of Education." *British Journal of Sociology of Education* 24, no. 5: 665–669.

——— (2005) "Global Inequality, Capabilities, Social Justice: The Millennium Development Goal for Gender Equality in Education." *International Journal of Education and Development* 25, no. 2: 111–122.

Walker, Melanie. (2005) "The Capability Approach and Education." *Educational Action Research* 13, no. 1: 105–122.

——— (2003) "Framing Social Justice in Education: What Does the Capabilities Approach Have to Offer?" *British Journal of Educational Studies* 51, no. 2: 168–187.

——— (2006) "Towards a Capability-Based Theory of Social Justice in Education." *Journal of Education Policy* 21, no. 2: 163–185.

Watts, M., and D. Bridges (2006) "The Value of Non-Participation in Higher Education." *Journal of Education Policy* 21, no. 3: 267–290.

Articles in Professional Journals, Newspapers, Newsletters, and Other Publications

Flores-Crespo, Pedro, and Mathias Nebel (2005) "Education and Development: Renovated Interpretations." *Maitreyee* 3 (October): 4–6.

Florian, Lani (2005) "Inclusion, 'Special Needs' and the Search for New Understanding." *Support for Learning* 20, no. 2: 96–98.

Nussbaum, Martha (2003)"Women's Education is Worth the Price." *Financial Times,* January 6. http://www.law.uchicago.edu/news/nussbaum-advocates.html.

Unterhalter, Elaine (2005) "Mobilisation, Meanings and Measures: Reflections on MDG3 and Girls' Education." *Maitreyee* 2 (June): 3–5.

Walker, Melanie (2005) "The Capability Approach and Social Justice in Education." *Maitreyee* 3 (October): 2–4.

Reports

Australian Government–Department of Education, Science and Training *Rethinking National Curriculum Collaboration: Towards an Australian Curriculum, Executive Summary.* http://www.dest.gov.au/sectors/school_education/programmes_funding/programme_categories/key_priorities/rethinking_national_curriculum (accessed July 22, 2005).

Unterhalter, Elaine (2003) "Education, Capabilities and Social Justice." Paper commissioned by UNESCO for the EFA (Education for All) Monitoring Report 2003. http://portal.unesco.org/education/en/file_download.php/0b4632378f877b196bbe66925d7d59f9Education,+capabilities+and+social+justice.doc (accessed March 3, 2003).

Unterhalter, Elaine, Rajee Rajagopalan, and Chloe Challender (2005) "A Scorecard on Gender Equality and Girls' Education in Asia 1990—2000." Bangkok: UNESCO.

Watts, Michael, Barbara Ridley, and Maggie Teggin (2005) *Expressing Dis/Abilities: An Evaluation of the Drake Music Project.* Norwich: University of East Anglia.

Conference Papers

Andresen, Sabine, Hans-Uwe Otto, Holger Ziegler (2006) "Education and Welfare: A Pedagogical Perspective on the Capability Approach." International Conference of the Human Development and Capability Association: Freedom and Justice, August 29–September 1, Groningen, the Netherlands.

Belisario, José, Monica Boross, and Virginia Schall (2005) "New Possibilities for Inclusive Education with Capability Approach." Fifth International Conference on the Capability Approach, September 11–14, Paris.

Biggeri, Mario (2003) "Children, Child Labour and the Human Capability Approach." Third International Conference on the Capability Approach: From Sustainable Development to Sustainable Freedom, September 7–9, Pavia. http://cfs.unipv.it/sen/program.htm.

———— (2005) "Education and Other Child Capabilities versus Child Work: Towards a New Definition of Child Labour Based on the Capability Approach." Fifth International Conference on the Capability Approach, September 11–14, Paris.

Biggeri, Mario, Renato Libanora, Stefano Mariani, and Leonardo Menchini (2004) "Children Establishing Their Capabilities: Preliminary Results of the Survey during the First Children's World Congress on Child Labour." Fourth International Conference on the Capability Approach, September 5–7, Pavia. http://cfs.unipv.it/ca2004/program.htm.

Boni, Alejandra, and Lozano, J. F. (2005) "Developing General Competencies through Ethical Learning to Increase Capabilities of University Students." Fifth International Conference on the Capability Approach, September 11–14, Paris.

Bridges, David (2006) "Adaptive Preference, Justice and Identity in the Context of Widening Participation in Higher Education." International Network of Philosophers of Education (INPE) Conference, August 3–6, University of Malta.

Brighouse, Harry, and Elaine Unterhalter (2006) "A Metric for Distributing Education: Primary Goods or Capabilities?" International Conference of the Human Development and Capability Association: Freedom and Justice, September, Groningen, the Netherlands.

———— (2002) "Primary Goods, Capabilities and the Millennium Development Target for Gender Equity in Education." Second Capabilities Approach Conference, September 9–10, Cambridge. http://www.st-edmunds.cam.ac.uk/vhi/nussbaum.

Burchi, Francesco (2006) "Education, Human Development, and Food Security in Rural Areas: Assessing Causalities." International Conference of the Human Development and Capability Association: Freedom and Justice, September, Groningen, the Netherlands.

Comim, Flavio, and Bagolin, Izete (2005) "Measuring Children's Education Capabilities." Fifth International Conference on the Capability Approach, September 11–14, Paris.

Flores-Crespo, Pedro (2001) "Sen's Human Capabilities Approach and Higher Education in Mexico: The Case of the Technological University of Tula." Conference on Justice and Poverty: Examining Sen's Human Capabilities Approach, June 5–7, Cambridge.

———— (2004) "Situating Education in the Capability Approach." Fourth International Conference on the Capability Approach, September 5–7, Pavia. http://cfs.unipv.it/ca2004/program.htm.

Flores-Crespo, Pedro, and Mathias Nebel (2005) "Identity, Education and Capabilities." Fifth International Conference on the Capability Approach, September 11–14, Paris.

George, Erika (2004) "Human rights, Development and the Politics of Gender Based Violence in Schools: Enhancing Girls' Education and the Capabilities Approach." Fourth International Conference on the Capability Approach, September 5–7, Pavia.

Hayes, M. T. and Hudson, R. (2005) "Guatemalan Schooling in the Global Context: Teachers' NO Concerns about Privatization and Neoliberal Development

Policies." Fifth International Conference on the Capability Approach, September 11–14, Paris.

Hinchliffe, Geoff (2006) "Three Dimensions of Capability." International Conference of the Human Development and Capability Association: Freedom and Justice, September, Groningen, the Netherlands.

Hodgett, Susan (2005) "Expanding Participatory Capabilities in Northern Ireland." Fifth International Conference on the Capability Approach, September 11–14, Paris.

———— (2006) "Expanding Participatory Capabilities in Northern Ireland: Sen, Education and the Pursuit of Eudaimonia." NO Paper presented to the Capabilities and Education Thematic Network, January 27, St. Edmund's College, Cambridge University.

Hoffmann, Anna Maria, Jan Van-Ravens, and Parul Bakhshi (2004) "Monitoring EFA from a Capabilities Perspective: A Life Skills Approach to Quality Education." Fourth International Conference on the Capability Approach, September 5–7, Pavia. http://cfs.unipv.it/ca2004/program.htm.

Lanzi, Diego (2004) "Capabilities, Human Capital and Education." Fourth International Conference on the Capability Approach, September 5–7, Pavia. http://cfs.unipv.it/ca2004/program.htm.

Lozano, Félix (2006) "Expanding Freedoms and Consolidating Social Justice: The Contribution from Higher Education." International Conference of the Human Development and Capability Association: Freedom and Justice, September, Groningen, the Netherlands.

Majumder, Amlan (2006) "The State and Plight of Indian Women: A Multidimensional Assessment of Well-Being Based on Sen's Functioning Approach." International Conference of the Human Development and Capability Association: Freedom and Justice, September, Groningen, the Netherlands.

Okkolin, Mari-Anne (2006) "What Do We Know and What Could We Know about Gender and Education in Tanzania." International Conference of the Human Development and Capability Association: Freedom and Justice, September, Groningen, the Netherlands.

Pérez Muñoz, Cristian (2006) "Examining the Universal Basic Income Proposal through the Capabilities Approach. Which is the Role of the Education in this Distributive Plan?" International Conference of the Human Development and Capability Association: Freedom and Justice, September, Groningen, the Netherlands.

Radja, Katia, Anna Maria Hoffman, and Parul Bakhshi (2003) "Education and the Capabilities Approach: Life Skills Education as a Bridge to Human Capabilities." Third International Conference on the Capability Approach, September 7–9, Pavia. http://cfs.unipv.it/sen/program.htm.

Raynor, Janet (2004) "Capabilities and Girls' Education in Bangladesh." Capability and Education Thematic Group Meeting, November, Cambridge.

Ridley, Barbara, and Michael Watts (2004) "Expressing Dis/Ability: The Drake Music Project." Paper presented at the British Educational Research Association (BERA) Conference , September, Manchester.

Sharp, J., and M. Watts (2004) "Go Tell It on the Mountain: The Value of RE beyond School. Paper presented at the British Educational Research Association (BERA) Conference, September, Manchester.

Suransky, Caroline, and Henk Manschot (2006) "Conceptualising a Pedagogy of Capability." International Conference of the Human Development and Capability Association: Freedom and Justice, September, Groningen, the Netherlands.

Terzi, Lorella (2004) "Beyond the Dilemma of Difference: The Capability Approach to Disability and Special Educational Needs." Third Anglo-American Symposium on Special Education and School Reform, June, Cambridge, and at the BERA Annual Conference, September, Manchester. http://k1.ioe.ac.uk/pesgb/x/Terzi.pdf.

——— (2003) "A Capability Perspective on Impairment, Disability and Special Needs: Towards Social Justice in Education." Third International Conference on the Capability Approach, September 7–9, Pavia. http://cfs.unipv.it/ sen/program.htm

——— (2005) "Equality, Capability and Social Justice in Education: Towards a Principled Framework for a Just Distribution of Educational Resources to Disabled Learners." Fifth International Conference on the Capability Approach, September 11–14 , Paris.

——— (2004) "On Education as a Basic Capability." Fourth International Conference on the Capability Approach, September 5–7, Pavia.

Unterhalter, Elaine (2004a) "Gender Equality and Education in South Africa: Measurements, Scores and Strategies." HSRC/British Council Colloquium on Gender Equity in Education, Cape Town. http://www.ioe.ac.uk/schools/efps/GenderEducDev/Elaine%20Cape%20Town%20paper%2015%20June%2004.doc.

——— (2004b) "Gender, Schooling and Global Social Justice." Fourth International Conference on the Capability Approach, September 5–7, Pavia.

——— (2006) "Global Inequalities in Girls' and Women's Education: How Can We Measure Progress?" International Conference of the Human Development and Capability Association: Freedom and Justice, September, Groningen, the Netherlands.

——— (2005) "The Political Foundations of Gender Equality in Education: Needs, Rights and Capabilities." Fifth International Conference on the Capability Approach, September 11–14, Paris.

Unterhalter, Elaine, and Harry Brighouse (2003) "Distribution of What? How Will We Know If We Have Achieved Education for All by 2015?" Third International Conference on the Capability Approach, September 7–9, Pavia. http://cfs.unipv.it/sen/program.htm.

Unterhalter, Elaine, Emily Kioko, Rob Pattman, Rajee Rajagopalan, and Fatmatta N'jai (2004) "Scaling up Girls' Education: Towards a Scorecard for Commonwealth Countries in Africa." Conference on "Scaling Up Girls' Education in Africa," February 2–3, Nairobi. www.ioe.ac.uk/efps/beyondaccess.

Vaughan, Rosie (2005) "Freedom through Education: Conceptualising Capabilities in Girls' Education." Fifth International Conference on the Capability Approach, September 11–14, Paris.

Walker, Melanie (2004) "The Capability Approach and Girls' Narratives of Life and Schooling in South Africa." Fourth International Conference on the Capability Approach, September 5–7, Pavia. http://cfs.unipv.it/ca2004/program.htm.

——— (2002) "Gender Justice, Knowledge and Research: A Perspective from Education on Nussbaum's Capabilities Approach." Conference on Promoting Women's Capabilities: Examining Nussbaum's Capabilities Approach, September 9–10, Cambridge. www.st-edmunds.cam.ac.uk/csc/conferences.

——— (2004) "Life Narratives of South African Undergraduates." American Educational Research Association Annual Meeting, April 12–16, San Diego.

——— (2006) "Towards a Mild Perfectionism and a 'Thicker' Human Capability Approach for Educational Praxis." International Conference of the Human Development and Capability Association: Freedom and Justice, September, Groningen, the Netherlands.

——— (2005) "What Are We Distributing in Education for Gender and Social Justice?" Fifth International Conference on the Capability Approach, September 11–14, Paris.

——— (2004) "What Insights Does the Capability Approach Offer to Education?" Symposium on the Capability Approach and Education, Australian Association for Research in Education, December, Melbourne.

Watts, Michael (2006) "Disrupting and Reproducing Social Injustice Through Part-Time Higher Education." Paper presented at the British Educational Research Association (BERA) Annual Conference, September, Warwick.

——— (2005) "Identity, Capability and Widening Participation at the Universities of Oxford and Cambridge." Identity and Capability Conference, Cambridge.

——— (2006) "Meeting the Needs of Refugees and Asylum Seekers with Higher Level Skills and Qualifications." Paper presented at the Conference on Developing Policy and Provision in Lifelong Learning with Refugees and Asylum Seekers in the UK, October 17–18, Cambridge.

——— (2004) "Sen and the Art of Motorcycle Maintenance." British Educational Research Association Conference (BERA), September, Manchester.

——— (2004) "The Tensions between Well-being and Adaptive Preference in Post-Compulsory Education in England." Conference on Capability and Happiness, June 16–18, Cambridge.

——— (2005) "What Is Wrong with Widening Participation in Higher Education." Conference on "Power, Discourse and Resistance," March 21–23, Plymouth, and at the Capability and Education Thematic Group Meeting, May, Cambridge.

——— (Forthcoming 2007) "Narrative Research, Narrative Capital and Narrative Capability." Paper to be presented at the Discourse, Power, Resistance Conference, Manchester Metropolitan University, March.

Young, Marion (2006) "Defining Valued Learning and Capability." International Conference of the Human Development and Capability Association: Freedom and Justice, September, Groningen, the Netherlands.

Young, Michael (2005) "Local Perspectives on Valued Learning Outcomes and Capabilities." Cortona Colloquium on Multi-Voiced Dialogue on Global Society, August 20–21, Cortona.

Theses and Unpublished Manuscripts

Flores-Crespo, Pedro (2002) "An Analysis of the Relationship between Higher Education and Development by Applying Sen's Human Capabilities Approach: The Case of Three Technological Universities in Mexico." PhD thesis, Department of Politics, University of York, England.

Hodgett, Susan (2004) "Government and the Making of Community: Policy and Social Development in a Globalizing World." PhD thesis, School of Education, University of Sheffield.

Okada, Yuko (2003) "The Capability Approach and Women's Literacy in India: A Critical Review." MA diss., Institute of Education, University of London.

Page, Elspeth (2005) "Gender and the Construction of Identities in Indian Elementary Education." PhD thesis, Institute of Education, University of London. http://www.elspethpage.freeuk.com.

Tanaka, Akiko (2003) "Can the Capability Approach Overcome the Problems of the Empowerment Approach? From the Perspectives of Women's Literacy." MA diss., Institute of Education, University of London.

Terzi, Lorella (2005) "Equality, Capability and Social Justice in Education: Re-Examining Disability and Special Educational Needs." PhD thesis, Institute of Education, University of London.

Curriculum Development and Learning Materials

Unterhalter, Elaine (2004) *Classroom Challenge: The Education for All Game*. A computer-based learning resource to assess countries' decision making relating to meeting the Millennium Development Goal for EFA by 2015 (with Joseph Crawford).

Index

Page numbers in italics refer to figures and tables.

Lightning Source UK Ltd.
Milton Keynes UK
UKOW050612070612

193928UK00001B/11/P